Biohistory

Decline and Fall of the West

Biohistory

Decline and Fall of the West

By

Jim Penman

**Cambridge
Scholars**
Publishing

Biohistory: Decline and Fall of the West

By Jim Penman

This book first published 2015

Cambridge Scholars Publishing

Lady Stephenson Library, Newcastle upon Tyne, NE6 2PA, UK

British Library Cataloguing in Publication Data
A catalogue record for this book is available from the British Library

ISBN (10): 1-4438-7130-3
ISBN (13): 978-1-4438-7130-3

TABLE OF CONTENTS

ACKNOWLEDGEMENTS

I offer my sincere thanks to Professor Paolini and his team, to LaTrobe University, RMIT and the Florey Institute for access to their laboratories and equipment, and to the Australian Research Council for providing matching funds. Neither the researchers nor these Institutions had any knowledge of the theory and ideas presented in this book. The conclusions and the interpretations are wholly mine.

My thanks also to Professor Ricardo Duchesne for his unstinting support and assistance with many aspects of the theory, to Dr Andrew Fear of Manchester University for his help with the section on Roman history, to Professor Michael McGuire for advice on primatology and general editing for sense and purpose, to Dr Frank Salter for his advice and valuable connections, and to Reece Russell for his work as a research assistant.

A special thanks to my son Andrew for his support over the years and as editor and co-writer. Without his contributions this book would not have been written, or at least delayed for many years. And last but not least to my wonderful wife Li. God grants a man no greater blessing than a loving and supportive wife.

INTRODUCTION

"It is better to light one candle than to sit and curse the darkness"
—Chinese proverb

At the age of 14 a friend gave me a copy of Thucydides' *The Peloponnesian War*.

Thucydides lived in Athens in the late fifth century B.C. He took part in the early stages of the great struggle against Sparta and was exiled, giving him the leisure to write his epic history of the war. In his day, Athens contained more men of genius than any city or nation in the history of the world. Within a period of no more than one hundred years, scarcely more than the lifetime of a man, Athenians took giant strides in such areas as philosophy, architecture, drama, sculpture, science, democracy and of course history. And yet, this great people were shortly to be crushed by Sparta and grow feeble.

Thucydides loved his city though fully aware of its faults, and his account was indescribably moving. Some years later when I read the book again I was unable to even read the section concerning the ill-fated expedition against Sicily. From that moment I began to ask the question "why?" Why did Athens fall from such brilliance to defeat and insignificance? And then came other questions.

Why did the Roman Empire collapse so completely, only a few centuries later? Why did Chinese dynasties so regularly dissolve into anarchy? Why are powerful and wealthy civilizations so often overrun by crude barbarians? So, as a nerdy and introverted teenager, I buried my head in the distant past, searching for answers.

None of the explanations I found made sense—they seemed to focus mainly on institutional changes or the decisions of leaders. But how could imperial policy cause people to have fewer children, or to refuse to join the army, or to abandon the use of money? Government efforts to reverse social trends were extraordinarily futile, as they have proved in our own day. It seemed that *people* were changing, but I did not know how or why. And in the collapse of Rome I found disturbing parallels with current events in the West. I can remember sitting in my school assembly hall watching the moon landings in July 1969, amidst all the excitement and hope for the future, feeling an ominous sense of foreboding. Could we be travelling down the same route as Rome? And if so, why?

At university I began to read widely over the whole range of human history, looking for patterns of behavior that could be linked to demographic changes and especially to population decline. I also found similar patterns of behavior in cross-cultural anthropology. And then, finally, in physiology and animal behavior. Could the key to history be not economics or politics but *biology*?

All of this was not a good move, career-wise. I had been warned, more than once, that the key to an academic career was to become a specialist in one small field. Now my field had become so broad that it not only covered all of history but anthropology and psychology and economics. And now I wasn't even prepared to stick to human beings as a species. Any academic position was out of the question.

But this was not the only problem. I had come to the conclusion that attitudes and behavior were very strongly influenced by early life experience, and that these effects were somehow physiological. To take this further I needed a biological research program, and this would require a great deal of money. So I set about turning my part time student lawn-mowing business into what became Australia's largest franchise network, with 3,300 Franchisees in four countries. In odd days snatched from the business I continued with my own research.

Finally, in early 2007, I approached Dr (now Professor) Tony Paolini in the Psychology Department at LaTrobe University, Melbourne, who had expertise in areas relevant to the theory. I offered funding to do some very specific experiments, mostly involving mild food restriction in rats. The results confirmed certain aspects of the theory and also helped to develop it further. These findings, together with recent breakthroughs in the new field of epigenetics, provided a biological explanation for the historical and demographic patterns observed.

A note on evidence

This book fully explains biohistory and is complete in itself, but only contains some of the supporting evidence, especially in relation to the biological sciences. Readers interested in a fuller picture, including how biohistory applies to different historical periods and full academic references, are referred to my book *Biohistory*.[1] Details are available at www.biohistory.org.

CHAPTER ONE

OF SCIENCE AND TEMPERAMENT

Most people say that it is the intellect which makes a great scientist. They are wrong: it is character.
—Albert Einstein

In 2007 the collapse of the U.S. housing market plunged the financial world into crisis. Trillions of dollars had been invested in mortgages with poor security, which was laid bare by the fall in house prices. Many mortgage lenders went bankrupt. Major institutions such as Northern Rock, Bear Stearns and Fannie Mae were taken over or nationalized to prevent a wholesale meltdown of the financial system. Western economies were plunged into recession.

Governments used all the levers that economic theory said would solve the problem. Deposits were guaranteed, economies primed with massive government spending, and interest rates reduced to near zero. Then they sat back and waited for the recovery that must surely come.

Seven years later, for much of the developed world, it has yet to arrive. Growth rates are anaemic or even negative. Unemployment through much of Europe is at catastrophic levels, especially among the young. Government debt has spiralled out of control. Greece is effectively bankrupt and other countries are on the edge, torn between unsupportable debt and the fear that further austerity might cause an outright collapse.

America is doing better, but even here there are ominous signs which long predated the crash. Real wages more than tripled between 1875 and 1975 but have been largely stagnant ever since.[2] Birth rates have plunged below replacement levels in all Western nations, with the consequent prospect of declining, aging populations. People are losing faith in government. Fewer of them vote, and membership of political parties is at a fraction of its former levels. The gap between rich and poor has grown dramatically, with a hollowing out of middle income earners.

It is not only economic and political indicators that are deteriorating. Obesity levels are rising and drug use is epidemic among the young. Sperm counts and testosterone are falling, and there are ominous signs of a rise in infectious disease.

Parallels can be seen in the history of ancient Rome. In that time there was also a growing gap between rich and poor, with sturdy peasant farmers giving way to vast slave estates owned by wealthy aristocrats. Faith in government collapsed, leading to the end of republican rule. The birth rate plummeted. The economy went into a long-term decline, from which it never recovered. It is worth noting that these trends occurred in ancient Rome—as in the modern West—after society had begun to cast aside its traditional religious and moral systems, especially those relating to control of sexual behavior.

Such parallels are only useful to us, of course, if we know *why* the Roman Empire collapsed, because only then can we know whether the same forces are in action today. Biohistory provides a clear answer to this question, and also makes clear that the same thing is happening today and for exactly the same reasons. It also explains why the decline was briefly checked in the late third century AD, and why the Eastern Empire did *not* collapse in the fifth century

But there is more to biohistory than just the decline of civilizations. It also explains how and why civilizations arise. It casts light on why the Industrial Revolution took off first in northern Europe, and why Japan, uniquely among non-Western nations, was able to swiftly adopt and use the new industrial technologies. It also helps to explain why most of Africa, despite almost a century of aid, remains desperately poor and backward. It takes particular issue with the idea that this might be about race, or genetic differences.

Biohistory proposes that the key to all of this—from the decline of Rome to the Industrial Revolution and the current financial crisis—is temperament. Some countries are wealthier than others because the people in them are harder working, more innovative, more willing to sacrifice present consumption for future benefit, less inclined to corruption as a government official, and so forth.

This is not a moral judgment. Wealthier peoples may also be less generous to friends and family, less indulgent with their children, less spontaneous, and greedier. Nor does it mean that all people in the society fit some national stereotype. For example, some people in society A may be harder working than many in society B. But if the average citizen of society A is harder working than the average for society B, this may have profound implications for wealth and other characteristics of each.

This is not a unique insight. In his superb book *A Farewell to Alms*, economic historian Greg Clark shows how the temperament of English people changed since the Middle Ages, such as in their working longer hours and being more prepared to sacrifice present consumption for future benefit. One example is the increased price of land relative to rental return, which meant people were prepared to accept a lesser return on their investment. He maintains that this change fully explains the economic explosion of the late eighteenth and early nineteenth centuries.[3] He does not provide an explanation for the change, apart from a suggestion that it may be genetic, but his evidence that there *was* such a change is powerful and convincing.

Temperament can also be used to explain political and institutional changes. One of the key distinctions biohistory makes is between "personal" and "impersonal" loyalties. The strength of political leaders ultimately depends on who supports them and to what extent. When loyalties are at their most personal, people will only support a leader they know well. At one extreme this means that political power cannot extend reliably beyond the local village, since a local leader can always prevail over one from the neighboring village. At most a leader can drive away the enemy and take their women and land, but as a section of the community takes over the vacant territory it becomes politically independent.

As loyalties become more impersonal they can extend to a local baron or tribal leader, who might be seen occasionally but is less well known. The next step is a king, rarely seen but still an identifiable individual. The most impersonal loyalties of all are to the laws and institutions of a republic.

As an illustration, consider the career of Richard Neville, Earl of Warwick in fifteenth-century England. Originally a supporter of King Henry VI, he became the chief supporter of the house of York and helped to put King Edward IV on the throne. Finding his influence curbed by the queen's family, he switched sides again and helped restore Henry VI, before being defeated and killed in a final battle which brought Edward once more to power. His followers seem to have simply gone along with all these changes, fighting for and against whichever claimant their lord told them to. Their loyalty was personal and local to their lord, whom they knew, rather than to their king.

To use a modern analogy, if the governor of California tried to depose Barack Obama and make Mitt Romney President he would gain very little support. Even soldiers and policemen who had voted for Romney would most likely ignore or arrest him, because their loyalty to the Constitution would outweigh their support for the man. In fact, in the present political climate such an attempt would be so futile as to be considered evidence of insanity. Six hundred years ago, this was politics as usual.

The same principle of changing temperament can explain the decline of Rome. As will be shown in chapter twelve, there was a clear change in the character of the Roman people during the late Republic and early Empire. As loyalties became more personal the Republic gave way to the Empire, and as they became more personal still the Empire itself collapsed. At the same time, an advanced market economy (which is an impersonal way of exchanging goods) changed to one based on subsistence farming and tributes to local leaders.

More recent events can also be explained in these terms. Saddam Hussein, as ruler of Iraq, was a brutal tyrant. When his health minister merely advised that he step down temporarily to help peace negotiations with Iran, the minister was sacked, arrested and killed, and pieces of his dismembered body delivered to his wife the following day. Saddam's campaigns against rebels and regime opponents involved poison gas, torture, assassination and (according to Human Rights Watch) the estimated loss of 250,000 lives.[4] Many more died in his abortive invasion of Iran. Many, if not most, Iraqis lived in terror of the regime.

In March 2003, the United States and its allies invaded Iraq, aiming to depose Saddam Hussein and thus bring peace and democracy. More than ten years later, with a trillion dollars spent and countless lives lost, they withdrew without having achieved either goal. The new government proved hardly more democratic than the old one, and was menaced by a brutal new foe in the Islamic State.

The answer to the puzzle of why Iraq did not become a peaceful democracy can be found in a community study done fifty years ago in Egypt, another Arab country with a very similar culture. The people of Egypt tended only to accept authority that was harsh and intimidating, indicating a fundamentally different temperament to that of people in the democratic nations of the West.

> The people thought of authority as necessarily involving an assertion of power and dominance, and could not respect those who did not display these attributes. Writing of the eighteenth century it was observed that, "if the peasants were administered by a compassionate multazin they despised him and his agents, delayed payment of taxes, called him by feminine names ... They still consider both Government and Government officials as agencies of imposition and control, and hence to be feared."[5]

When people only obey rulers who are brutal and terrifying, it is brutal and terrifying men who make the most effective rulers. More lenient men are ignored or brushed aside. The United States and its allies thought that removing Saddam would turn Iraq into a peaceful democracy. They were wrong, because Saddam's rule merely reflected the kind of leadership the majority of Iraqis were temperamentally inclined to accept.

The same can be said of nations affected by the Arab Spring in recent years—either ongoing chaos (Libya and Syria) or renewed autocracy (Egypt). Similar patterns occurred after the Russian Revolution of 1917 or the French Revolution of 1789. Ending one autocracy quickly gave rise to another one.

The idea that economic and political systems reflect the prevailing temperament is not conventional wisdom, but it is not original either. Nor is it original that such attitudes might have a physiological basis. Recent research has indicated that attitudes towards matters such as politics, religion, and capital punishment are deeply rooted in biology and that the 'reasons' given for them are largely rationalizations.[6]

Biohistory is the study of history, as well as of economics, psychology and anthropology, united by a common strand of evidence in biology. Different temperaments are traced back to the influence of early life, in particular the extent to which parents control or punish their children at different ages. For example, chapter five suggests that the classic Arab temperament stems from extreme indulgence of infants combined with harsh control of older children.

Biohistory takes issue with the idea that differences between peoples can be explained by genetics, such as the idea that Europeans and East Asians are more intelligent.[7] Even if such a difference could be demonstrated it would be far less important than differences in temperament determined by the environment. Overall, differences between and changes within societies cannot be explained in terms of inheritance. Genetically speaking, human beings are very similar to each other. It is often said that there is more genetic diversity in chimpanzees from a few hectares of rainforest than in the entire human race.[8] Genetic differences cannot explain, and are not needed to explain, differences in wealth, creativity, political institutions or much else that matters.

But at another level, people are profoundly different. This is at the level of epigenetics, the new science which looks at the way in which genes are switched on or off by the environment. Thus, two people with similar genes but different early environments can be remarkably different in attitudes and behavior, as different genes become more or less active. These epigenetic differences can make people more or less hard working, rigidly dogmatic or open to change, peaceful or violent, timid or forceful, honest or corrupt, accepting or rejecting of brutal authority, and much more. An example is given in Fig. 1.1 below.

Fig. 1.1. Example of an Epigenetic effect. A simplified overview of epigenetics, development and behavior. Early experience has a major influence on attitudes and behavior.

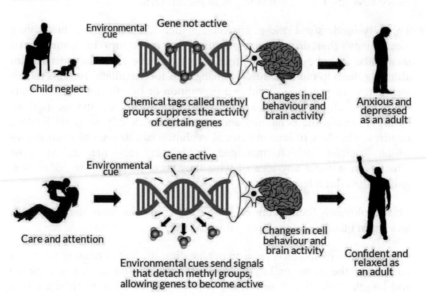

Environmental cue

Gene not active

Child neglect

Chemical tags called methyl groups suppress the activity of certain genes

Changes in cell behaviour and brain activity

Anxious and depressed as an adult

Environmental cue

Gene active

Care and attention

Environmental cues send signals that detach methyl groups, allowing genes to become active

Changes in cell behaviour and brain activity

Confident and relaxed as an adult

What is more, these differences tend to pass from generation to generation, partly by direct inheritance but more by the way children are treated in early life. And they have profound effects on the political and economic make-up of societies. If people are epigenetically primed to accept only the most brutal forms of authority, then governments will tend to be brutal or unstable. When people are epigenetically primed to be innovative, to act with integrity and inclined to work hard, national wealth grows. When men are epigenetically primed to be aggressive and proud, wars break out. Thus it is that biology, more than anything else, determines the nature of society.

Culture, the ideas and practices that define how people should think and behave, has a profound impact, but not in the way most people think. First, culture largely reflects the underlying character of the people. When people are aggressive by nature the culture is warlike. War is glorified, and men are praised and valued for courage and pride. But culture also has an impact on the underlying biology. Practices such as patriarchy, control of sexual behavior, religious rituals and different ways of rearing children all have epigenetic effects. These in turn cause changes to character, which in turn influence culture in an ongoing cycle. All of the questions given above have answers couched in physiological terms.

To fully understand these answers, which constitute the underlying mechanisms that drive human culture, we must turn to animals. All mammals, including human beings, appear to have an inbuilt mechanism allowing them to rapidly adjust to changes in food availability. This means they can change behavior within a generation or two to suit environments with chronically limited food or occasional famines. By a strange quirk of biology, these same behaviors and attitudes are exactly what civilization requires. The story of human cultural evolution can be seen as a process by which societies which managed to trigger this mechanism most effectively, without any idea of what they were doing or why, overcame those which did it less well.

This biological foundation of biohistory provides one major benefit lacking in other social theories—it is *testable*.

The scientific method has been an outstanding success in helping people to understand the world, and to develop technologies and drugs that improve and lengthen our lives. And at the core of the scientific method are two quite simple ideas. The first is that, all things being equal, we prefer the simplest theory to explain the available facts. And second is that a scientific theory should generate non-obvious hypotheses that can be tested, and on that basis the theory is confirmed, modified or refuted.

As an example, Einstein's theory of Relativity predicted that light should be affected by gravity and bent by a specified amount when passing near a massive object such as the sun. This had never been observed, and no competing theory made any such prediction. The trouble is that the sun is so bright that it drowned out light from distant stars. The only way to test the theory was by a total solar eclipse, observed in the right place and with exactly the right weather. Scientists spent many years traipsing around the world in pursuit of just such an event, and eventually made observations. The sun's gravity bent the light of distant stars, and by exactly the amount Einstein predicted. Thus was Relativity confirmed.

For the social sciences this approach has proved difficult, to put it mildly. To take just one example, historians have many different explanations for the Second World War, including the personality of Adolf Hitler, resentment at the Versailles treaty, aspects of German national character, and more. But the only way to absolutely prove any theory would be to run the twentieth century again without one such element (for example, take out Hitler), which is clearly impossible. By contrast, chapter nine explains war in terms of maternal anxiety, and suggests a form of blood testing that could confirm or refute such an idea.

Testing the theory

This is not a "common sense" view, but common sense is not a necessary criterion for a theory to be valid. For example, neither Relativity nor Quantum mechanics are especially plausible theories. Light can be "bent" by gravity? A particle can be in two places at once?

Biohistory is science in that it explains a wide range of data, and it is testable both inside and outside the laboratory. The research program cited earlier is an example of just such testing, as a result of which the theory has been confirmed in some areas and modified in others. It may be noted that biohistory is the only theory of history ever to have resulted in ten papers (and counting) in high ranked biomedical journals. Each chapter contains an example of proposed tests. It is my hope that researchers will take up the challenge and put biohistory to the test.

In the next chapter we will look at aspects of family and personal behavior that are associated with large political units and advanced economies. By finding those same characteristics in certain animal populations, and working out their physiological basis, we will begin to understand the biological foundation of the temperament that underpins civilization.

CHAPTER TWO

FOOD RESTRICTION

Every time you cuddle with your children, you're likely to be driving down your testosterone.
—Helen Fisher

This book is concerned with the mechanisms by which biology underpins human culture and civilization. In order to understand those mechanisms, we start by contrasting the behavior of civilized peoples with those in smaller scale or tribal societies. Some of these differences are obvious. Civilized peoples are more likely to form large states and obey distant leaders they have never met. They are more likely to trade goods and use money rather than operate a subsistence economy. And compared with hunter-gatherers or primitive horticulturalists they are generally better at the routine work of farm, factory and office.

But looking more closely it becomes clear that there are also surprising differences in family and personal behaviors. Compared to small-scale societies, civilized peoples are far more likely to systematically control and direct their children's behavior, to form nuclear monogamous families, and to delay the start of sexual activity. [9]

An example of one such a community is the Japanese village Niiike, studied in the 1950s. Niiike was part of a nation state which controlled tens of millions of people and had an advanced industrial economy. Control of children was systematic and consistent from earliest childhood. [10]

> Training of children beyond the toddler stage is a conscious goal of Niiike parents and grandparents. They have certain well-defined goals to which they direct their efforts ... The child must learn early and well, however, to obey and conform to certain inflexible rules. Obedience is easier, perhaps, in that these are rules for a way of life followed by everyone he knows, not rules made especially for children. Many express the pattern of hierarchy. Speech is one example.

Their family and sexual patterns were those which cross-cultural analysis shows to be typical of civilized societies. All marriages were monogamous, and brothers who married moved out of the family household to form new nuclear families, though one son (usually the eldest) continued to live with and look after his parents. There was little premarital sex, at least for women, and the stigma of sexual misbehavior made it much harder for a girl to marry. Marriages were arranged at a relatively late age, and marital sex did not seem to be very rewarding for women.[11]

> The sex act itself usually is a brief, businesslike affair with a minimum of foreplay. The husband, after waiting in the quilts at night for the rest of the household to settle into slumber, grasps his wife and satisfies himself as quietly and inconspicuously as possible.[12]

This same pattern has been the norm in all Western societies until quite recently, and in a slightly different form throughout India, China and the Middle East.

Consider by contrast the Yanomamo, an indigenous people of the Amazon rain forest. They had no political organization beyond the village, and even villages were commonly split by vicious feuds. Much of their food was gained by horticulture but the work only took a few hours a day, and the men tended to spend more time hunting.

Though this was a fiercely warlike people their attitude to children was lenient in the extreme.

> Yanomamo are indulgent with children ... Children are punished infrequently. However, a severe beating is sometimes given suddenly by an angry parent. Spanking, or other formalized punishments, are not used.[13]

Boys in particular enjoyed a relatively control-free childhood that could stretch into their late teenage years, compared to girls who started work much earlier by assisting their mothers with chores.

Marriages were polygynous or even polyandrous in form, with successful men typically having multiple wives. The less successful might have none, or a number of men might share a single woman. Sexual activity started early and married women frequently had affairs, despite the brutal punishments inflicted by jealous husbands. The abduction of women was a common source of feuds. A bigger contrast to the Japanese pattern could hardly be imagined.

Note that the key distinction in terms of childrearing is in *control*, rather than punishment. Although Yanomamo children were allowed more latitude and were indulged, Japanese parents were actually *less* likely than the Yanomamo to inflict painful punishments on their children. Parents in civilized societies may or may not punish their children, but they almost always control them, giving them fewer liberties and more prescribed rules of behavior.

Given the different levels of development between these two societies, it is important to understand the nature of these behavioral differences, what they mean, and where they come from.

Our first clue to understanding this puzzle comes from the study of animal societies. Curiously enough, all of the mating and childrearing behaviors that distinguish complex and small-scale human cultures have direct equivalents in monkeys and apes. As an example, some populations start to have sex and breed immediately after puberty, while others delay for many years. Some form troops where males have access to several females, while others defend territories as monogamous pairs. There is no exact equivalent to the control exercised by human parents, but some populations favor their young with far more time and attention.

The one common point about animals which exhibit behavioral traits that in humans are associated with civilization (late breeding, monogamous nuclear families, time spent with young) is that they tend to live in places where food is in short supply. This does not mean they are starving, but that they routinely experience mild hunger. Mild hunger is defined as equivalent to what humans experience during weight loss programs; there is no health-threatening malnutrition, just a diet comprising less food than one would like to have.

Chronic mild hunger gives rise to a series of hormonal, behavioral and epigenetic changes which adapt a species' behavior to a food-limited environment, enabling efficient use of available resources. By a strange quirk of biology, these same physiological changes have produced behavioral traits in human beings which, taken collectively, have adapted people to the requirements of civilization. They make people harder working, better at trade, more able to co-operate in large groups. For this to work, people do not even need to be hungry. Human societies, by a process of trial and error, have developed cultural practices which mimic the physiological effects of hunger. Thus we can act and think like hungry primates, even though we are not actually short of food.

It is variation in the level of this "hunger" mechanism which explains many differences between societies, especially economic success and type of political organization. Changes in the level of this mechanism also explain why civilizations rise and fall, as later chapters will explore.

Primate social behavior—gibbons and baboons

To get an idea of where these different behavioral effects come from and what their use in the wild might be, let us begin by looking at two species that live in very different environments: Asian gibbons and African baboons.[14] Each species is subject to very different patterns of food availability, as well as having to deal with differing sets of environmental hazards, such as predators. In consequence, they display markedly different social behaviors.

Gibbons are found in the tropical forests of Southeast Asia. Living high in the forest canopy, they are physically adapted for brachiating—swinging spectacularly from tree to tree with their immensely long arms. There are many different species of gibbon, grouped within the family Hylobatidae, but their behavior is quite consistent across species.

One of the remarkable things about gibbons is that they are some of the fussiest eaters on earth. They will only eat the leaves or fruits of certain trees, and then only at certain stages of ripeness. The peculiar consequence of this is that, even though they live in one of the world's lushest habitats, calories are hard to find. Their favored foods are widely and thinly scattered through the forest, so gibbons have to spend a lot of time foraging and traveling.

Despite this scarcity, gibbons rarely starve. There are food plants available in every season, and there is little variation from year to year. There is always something for them to feed on, no matter how difficult it may be to find. For most gibbon populations, being hungry is a fairly constant state. For Malaysian gibbons, food is so hard to find that individuals can only just maintain themselves, even with an exclusive territory. Squabbling within groups over food is common and is the main way in which young are expelled from a group. Because they live high in the trees, gibbons are rarely taken by predators. This means that gibbon populations are limited only by the availability of food.

The habitat of African baboons is very different from that of gibbons. They are ground-dwelling. Although they can be found in hilly country and woodland, the species we are interested in live on the open savannah. Unlike gibbons, baboons are omnivores and not overly picky about what they eat. Predators, mainly leopards and lions, are a constant hazard. Together, these two factors mean that there is usually plentiful food.

But the baboon environment is also much less stable. In times of drought, food can run low or even disappear, resulting in severe hunger and even starvation.[15] Thus, baboon populations are limited not by chronic food shortage but by predators and occasional starvation.

As you would expect, the social behavior patterns of baboons and gibbons are markedly different, in keeping with the habitats in which they live.

Gibbons, living in a state of constant low-level food restriction, are relatively unsociable. They tend to form very small social units, typically a mated male and female. Each pair lives within a territory which they defend from other gibbons. Adult males drive away other males, and females other females, including their own offspring once they pass the age of puberty. In biological terms, this makes very good sense. When food is limited, only an exclusive territory makes it possible to successfully rear young.

Baboon behavior is at the other extreme. Baboons spend far more time socializing within their group, which is much larger than that of gibbons. Their troops normally consist of multiple males and females (though in areas of more limited food there may only be one male with several females). The dominant male tends to monopolize females and sire most of the young, with other males having access only when he is distracted.[16]

Since predators are plentiful and hard to avoid in the open, baboons are active in defense. Against lions they raise the alarm and attempt to hide in trees (if available), but a leopard will be fiercely mobbed if cornered in a tree or a hole.[17] It is not uncommon for leopards to be killed in such attacks. This behavior also makes biological sense. With food plentiful there is no need for an exclusive territory, and a large troop can give warning against lions and help mob leopards.

By contrast, gibbons are timid in the face of threats, and will swing away through the treetops at the slightest danger.[18] This again makes biological sense—there is no point risking your life by attacking a predator that is unlikely to catch you.

Gibbons and baboons differ in their reproductive strategies. Gibbons are slow breeders. They are long-lived, discriminating in their choice of mates, and breed very slowly—much more slowly, in fact, than they physically could, as pairing and reproduction are delayed until well after sexual maturity. [19] Puberty is also later than for other species. This makes perfect sense in an environment with scarce food, where too many offspring could either put disastrous pressure on resources or result in very high infant mortality (which is a waste of the energy resources put into breeding). The dangers of overly fast breeding are illustrated by the fact that gibbons are prone to miscarriage and premature birth, all factors likely to result from poor nutrition. [20]

Baboons breed early and fast and are less choosy about mates. With food plentiful most of the time, the young are more likely to survive. Fast breeding also compensates for deaths caused by predation.

There are other differences. Baboons tend to wean early and provide little care after that age. Gibbons keep their young close until puberty. Baboons find food easily and spend much of their time socializing and resting. Gibbons spend far more time foraging for food, necessary when food is limited but also due to a matter of temperament (as confirmed by the rat experiments described below). In human terms they are primed to work hard, which is a key characteristic of people in civilized societies.

Changing food-restricted behavior

There is still another crucial difference between gibbon-type and baboon-type social behavior—adaptability. Gibbon behavior is adapted to a food-short environment but is not a direct response to it. For instance, even when gibbons have plentiful food (such as in captivity), they remain slow breeders and are socially intolerant. These behaviors are evidently rooted in their genes. [21]

Baboons are far more flexible. The environments they live in are more varied and include savannah, woodland and desert. Baboons in more stable environments with low predation (such as woodland) behave more like gibbons, forming smaller groups and breeding later and less frequently. It seems they can change their behavior as a direct response to food shortage.

It does not take much thought to realize that such flexibility provides a huge advantage. Animals entering an environment where food is short can quickly adapt to the new conditions, chasing away competitors and slowing breeding to maximize the survival of their young. But if enough are killed that food becomes more plentiful they can immediately breed faster to make up the numbers.

Such flexibility is common in primates. Vervet monkeys are like baboons in forming multi-male troops when food is plentiful, and one-male troops when it is more limited. Mentawaian langurs are bolder and form larger troops in regrowth areas with plentiful food, and monogamous nuclear families in untouched forests with more limited food.

It is important to note that this flexibility has limits. Baboons and vervets have not been observed to form monogamous nuclear families, and Mentawaian langurs vary from monogamous pairs to one-male troops, but do not form multi-male troops. Each species has a range of variation that goes only so far. Baboons and vervets on the "food-plentiful" end, and langurs on the "food-short" end. Gibbons are also on the food-short end, but with little variation. Each has a "set point" which determines the behavior that is most natural to the species and presumably set by genes, but with a limited ability to change in each direction.

Where do humans fit in? Judging by the behavior of hunter-gatherers, who are closest to our ancestral way of life, we belong on the "food plentiful" end. Hunter-gatherers normally live in multi-family groups which travel and hunt together. For example, the Mbuti pygmies, who live in the Congo region of Africa, form social units consisting of at least six to seven families, the minimum required for the Mbuti practice of hunting with nets. The maximum size of such groups is determined partly by the needs of hunting, with too many people seen as a disadvantage.[22] In short, this is nothing like the gibbon or langur pattern where couples or polygynous males defend an exclusive patch of land. In this respect human hunter-gatherers act more like baboons and vervets than gibbons. Humans are far more likely to form pair bonds than most primate species, but do so *within* the multi-male band. In other words, such pairs do not defend exclusive territories as gibbons do. A suggested "set point" and range of variation for each species is given in Fig. 2.1 below.

In overall behavior, the lifestyle of modern hunter-gatherers suggests that ancestral humans were more like group-living baboons than territorial gibbons, apart from unusually strong pair bonds. This mechanism, which adapts behavior and attitudes to the level of food availability, is the key idea of biohistory and the main driving force behind civilization. Given that civilized societies are more successful at producing food it might be expected that they would show more food-plentiful behavior. But, as we have seen, civilized peoples show all the characteristics of *food-restricted* behavior.

Fig. 2.1. Proposed set points and range of variation in food-restricted behavior in various primates including humans.

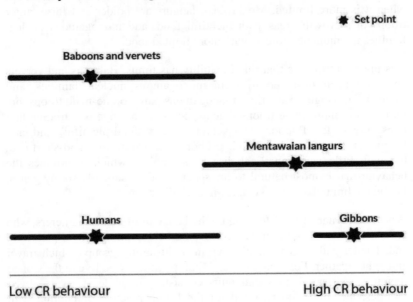

If this theory is to hold water we must answer one question—how does the mechanism work? How could mild hunger result in all these far-reaching changes in behavior? And given that, why do civilized peoples show similar characteristics when not especially short of food?

Food restriction—hormones, epigenetics and behavior

To study the food shortage mechanism in monkeys or humans would be expensive and very difficult. Fortunately, rats and mice demonstrate exactly the same responses. For example, a recent study supplemented the diet of mice with a little extra sugar—about equivalent to a human drinking three cans of soft drink a day. This is, in effect, a condition of super-abundant food. These mice not only had a higher death rate than the control mice, but 26% fewer males were able to establish territories.[23]

Thus, in 2007 I approached Dr Tony Paolini with a proposal to study the effects of mild food restriction on the physiology and behavior of laboratory rats. This is a subject surprisingly little studied in rats, much less than starvation. As indicated earlier, starvation or malnourishment has very different effects from mild restriction. It is hugely stressful and dangerous to health, compromising the immune system, causing extreme fatigue and weakness, irritability, anaemia, apathy, reduced coordination, and loss of concentration, as well as acute liver damage.

In our studies, rats were given food-restricted diets 25% below the level that would be taken by free feeding (in some experiments rising to a 50% reduction for short periods). The diet contained adequate levels of protein, vitamins and minerals, and was in no way detrimental to health. It was similar to that of properly conducted weight-loss programs for humans.

What follows is a short overview of the experiments, the results and comparable evidence. A fuller description of the experimental procedures and data can be found in *Biohistory*.[24]

Testosterone reduction

Previous studies of acute (i.e. short-term as opposed to chronic long-term) food restriction have found that it reduces the male sex hormone testosterone in a variety of mammals, including humans. Our studies confirm this, showing reduced testosterone in rats with mild chronic food restriction.

In other words, a *reduced diet brings down the level of testosterone*. The social implications of this are significant. High testosterone is associated with stronger sex drive, aggression and dominance. Men with high testosterone are less likely to be married or in a committed relationship, and when they are married they spend less time with their wives. If we think back to our gibbon and baboon groups, the variation of testosterone as a result of restricted or abundant food fits with the more active sexuality and weakened pair bonds of baboons (whose environment contains plentiful food most of the time), and the stronger mating bonds of gibbons.

The greater aggression of baboons in both their social behavior and their response to predators is also consistent with higher testosterone. The link between testosterone and aggression is strong, and has been found in both animals and humans. The most violent prison inmates (including women) have higher levels of testosterone. Men with high testosterone are more likely to be delinquents, to use drugs or abuse alcohol. And while they tend to be more effective as combat soldiers, they are more likely to go AWOL. Testosterone is also linked to sensation-seeking and high-risk behavior. High-testosterone people are also more gregarious, more likely to need the company of others, and are less happy alone. This too is consistent with the larger social groups formed by baboons.

Given the assertiveness of high-testosterone individuals and the tendency for the dominant males in animal groups to have high testosterone, it seems odd that status in human society (or at least modern Western societies) seems to be inversely related to testosterone. In our culture, high status tends to be linked to a person's profession, and occupationally successful people usually have *lower* testosterone. Testosterone levels tend to be highest in the unemployed, next in blue collar workers, lower in sales, and lowest in professional occupations.

Other occupations, not necessarily high status, have distinctly low testosterone. The lowest levels of all are found in farmers, and country people in general seem to be lower in testosterone than city dwellers. Ministers of religion also tend to have low testosterone.[25]

Reduced testosterone is also associated with more attentive maternal care. This fits with the closer care of infants observed in human societies in food-restricted environments.

Stress hormones

Hunger is a stressful experience so we might expect a restricted diet to increase stress. In fact, mild food restriction has a surprising effect on stress hormones. It raises the level of corticosterone, but reduces or has no effect on adrenocorticotropic hormone. This is an interesting result, because corticosterone (or its equivalent cortisol in humans) acts to minimize the harmful effects of stress. It also helps recovery from stress by winding down the stress response. People who have post-traumatic stress disorder show lower levels of cortisol than those who have recovered from their traumatic experience without developing PTSD.

High cortisol can also eliminate the aggressive effect of testosterone; adolescent males with high cortisol and high testosterone are no more aggressive than those with low testosterone. In fact, high corticosterone/cortisol actually lowers testosterone. This is another fact that will be of particular interest when we come to look at the early development of human civilizations.

Sex hormones and sexual behavior

A restricted diet has significant effects on fertility and sexual behavior. Mild food restriction in rats reduces luteinizing hormone (which stimulates ovulation in females) and follicle-stimulating hormone (which promotes sperm production in males).

Combined with the lowering of testosterone, this would lead us to expect food restriction to reduce sexual behavior. And indeed, this is what we find. Laboratory studies have found that food-restricted animals are less fertile, reach puberty later, and have lower rates of sexual intercourse.

Moderate reduction in food intake reduces sex drive in humans as well. Practitioners of low-calorie diets commonly report a reduction in libido and a loss of interest in sex in general.

Our study also found that food-restricted male rats were less attractive to females. Presented with males on a 50% food restriction, females spent less time in their vicinity, indicating reduced sexual interest. These findings are confirmed by studies showing that females of many species are much less interested in food-restricted males than in those who eat all they wish.

Studies of humans also show this pattern. Women who are in the most fertile period of their menstrual cycle tend to prefer the faces of men with higher levels of testosterone. Somehow, women are able to decipher men's testosterone levels from facial cues and thus make judgments on mating ability, just as the females of other species appear to get the same information from smell. High testosterone in animals is a sign that a male is likely to be better fed and socially dominant. The same pattern can be found, at least anecdotally, in the greater dating success in high school of "jocks" compared with "nerds," even though the latter may have far better career prospects. Football players tend to have relatively high testosterone while, as noted earlier, professionals and other white collar workers tend to test low.[26]

In terms of primate behavior a jock is more like a baboon, being aggressively sexed and often attractive to teenage girls. A nerd is like a gibbon, timid and backward but preparing to establish a "territory" (i.e. a high paying job) that will make him more attractive as a mate in the long-term. Considering the role of testosterone, this is more than just an analogy.

The conclusion we can take away from this is that food restriction affects the levels of several sex hormones, which has the effect of delaying sexual maturity and reducing sex drive and sexual behavior. This again matches the observation of gibbon and baboon mating behavior.

Food restriction, health and behavior

Mild food restriction can have a wide variety of effects on the well-being and behavior of animals and humans. Many studies have shown it to be good for health in that it increases lifespan, improves disease resistance and reduces age-related illness. In particular, it can reduce diabetes, cancer, cardiovascular disease and brain atrophy. All of this reflects a shift of bodily resources from fast reproduction to body maintenance and longevity, which is exactly what is required in an environment with stable but limited food.

In addition, food restriction has been shown to improve learning, memory and motor performance in aged rats and mice, and to some extent even in younger ones. This is also helpful in an environment such as a tropical forest where food is scattered over a wide area. If a particular tree provides fruit in one brief season of the year only, it is vital to remember that fact and to know where to find it.

Rats on a limited diet explore more actively. This is not simply a search for food since the behavior persists for up to ten days after unlimited food is restored. This behavior is consistent with the faster movement and exploration of animals in a food-limited environment, a crucial asset when food supplies are scarce and scattered. Searching for food, even when not hungry, is the rat equivalent of human work. But though more exploratory they are more wary of cat urine. Fear of predators is characteristic of gibbons and a behavior well suited to a food restricted environment.

Food shortage reduces sociability, and is part of the reason why primates form smaller troops when short of food. A group of rhesus monkeys experiencing a severe three-week food shortage became much less sociable, with less grooming, playing and fighting. Likewise, a study of men on a reduced diet for six months found them to be much less social than normal.

Food restriction and parenting

Of all the areas of social behavior, parenting is one of the most significant. The type of care (or lack of it) given to offspring is a key element in their socialization, as well as the temperament and survival of the species. Restricted diet has a significant effect here too—an effect which, as we will see later, has deep implications for how human societies have developed.

In general, animals in food-limited environments give their offspring more attentive care. Female rats spend more time close to their pups, nurse more of them and for more of the time, suckle them more intensively, and engage in a greater number of maternal activities than those given unlimited food. Our own study found that females given a reduced diet while nursing kept their nests in better condition and were quicker to retrieve their pups when scattered.[27] This is quite dramatic to watch. Whereas the fully-fed mothers were listless and incapable, the food-restricted mothers were efficient and capable. This is perhaps the closest rat equivalent to the control exercised by human parents, which as we have seen is characteristic of civilized peoples.

Reduced diet, maternal stress and the effects on offspring

As noted earlier, mild food restriction tends to reduce testosterone and the level of harmful stress hormones. In rats, hunger (and other stresses) experienced by a mother can also have a significant effect on these hormones in her offspring. Severe hunger raises the level of stress hormones in a mother's offspring, but mild hunger reduces them.

However, the effect on testosterone is very different. Mild hunger experienced by the individual reduces testosterone, but when experienced by a mother if anything it *raises* the testosterone in her offspring, especially when they have plentiful food after infancy. The young also tend to be more dominant and less fearful. In other words, these effects are in some ways *opposite* to the effect of food restriction on adults, a finding that will come to have major significance when applied to human societies. This is one aspect of a common finding that will be referred to on a number of occasions—environmental influences have very different effects when applied at different ages.

Food restriction and epigenetics

In the previous chapter we mentioned the burgeoning field of epigenetics, in which environmental influences turn up or down the activity of certain genes, while not altering the DNA. Our studies measured the activity level of a number of genes with known effects on hormones. They found that subjecting female rats to a reduced diet has profound epigenetic effects on their offspring, in some cases doubling and in others halving the genes' level of activity. Changes of this magnitude can easily account for the hormonal and behavioral changes observed.

One strong effect is on the gene encoding the androgen receptor, which regulates the production of testosterone. Like a thermostat, it reacts to a high level of testosterone by shutting down production and thus preventing it from rising past a certain level. Diet reduction in mothers causes a dampening of the activity of this gene in the offspring, which would explain their higher testosterone levels. By contrast, experience of mild hunger in later life *increases* the activity of this gene, which is also in keeping with the lower testosterone levels described earlier. In other words, for this and other genes the effects of early or maternal diet reduction were different and even opposite to those of diet reduction in later life.

Food restriction, culture and human societies

So what are the implications of these animal studies and biochemical experiments for the history of human societies? A number of key observations can be made. The behavior of animals in food-restricted environments is a direct response to food shortage. Evidence from primate studies, plus laboratory findings and medical research, shows that mammals develop such behaviors as a response to mild hunger, and these changed behaviors adapt them to a food-limited environment.

Further, each species has a genetically determined "set point" for this behavior, and a fixed degree of adaptability. In other words, the mechanism will only allow them to deviate a certain amount from their set point. Baboons, for instance, have a lower set point than gibbons (i.e. their level of restricted-food-type behavior is lower), but are much more adaptable around their set point.

The biological mechanism that underpins this—the effects of food reduction on hormones and behavior—will cast light on some otherwise puzzling and inexplicable features of human societies. For example, we have seen that people in more complex societies tend to show family and social behaviors characteristic of food shortage, even when food is plentiful. In the next chapter we will see how this is achieved, and that understanding this can explain the rise and fall of civilizations.

Testing

The effect of mild food shortage on behavior could be tested by setting up a sizeable population of animals in conditions as naturalistic as possible, with plentiful space and the ability to systematically control food levels. Expected changes from limited food would include larger and better guarded territories, better maternal care, more active exploration, and caution with respect to predator odors.

CHAPTER THREE

THE CIVILIZATION FACTOR

Science and religion are the two most powerful forces in the world. Having them at odds ... is not productive.
—E. O. Wilson

In the last chapter we saw how behavioral patterns such as nuclear monogamous families, slow breeding and "hard work" are characteristic of animals in food-restricted environments, and also of civilized human societies. We also saw that limited food tends to produce the behavior that helps animals prosper in these environments. This chapter will explain how human cultures, and especially religious systems, change temperament and behavior in a way that mimics the effects of food shortage—even when food is not especially short. We will also come to understand how such changes help societies to prosper and expand.

An example of the way temperament affects economic behavior can be found in the Mbuti pygmies, a tribe of nomadic hunters and gatherers in the Ituri rain forest as described earlier. In recent times, Bantu speaking farmers began colonizing parts of the forest, converting land for agriculture. The Mbuti appreciated the benefits of farming, but were far less keen on the hard physical work it demanded:

> The village was still a place where they could relax, enjoy luxury foods, wine and tobacco, in return for a minimum amount of service. Being a practical people, however, they responded, to a token extent, to the new demands for service, in the form of plantation labor. But it was not to their taste, and it proved highly detrimental to their health. The villagers in many cases realized this and compromised by using the Mbuti mainly to maintain guard over the plantations, so they would not be affected by the unaccustomed sunlight or exercise, and only in emergencies asked them to actually work in the fields.[28]

It is also possible for farmers to lose their taste for the hard work involved in agriculture. This occurred in the late Roman Republic, as will be discussed in chapter 12. The peasant farmers who had been the backbone of the army began flocking to the city, where they caused trouble and increasingly lived off government handouts. A group of progressive politicians led by the brothers Tiberius and Gaius Gracchus attempted to resolve the problem by giving farms to the urban poor, but this well-intentioned reform quickly ran into trouble. The new farmers quickly sold up and returned to their accustomed lifestyles in the city.[29] They no longer had the temperament for farming, and in this sense had become more like the Mbuti.

As we saw in the last chapter, farmers tend to have low testosterone, which in turn is associated with the physiological complex that stems from chronic mild hunger. Returning to our zoological foundations, gibbons must spend a great deal of time in a systematic search for food, which is more routine than socializing, fighting, hunting and other primate activities. In this sense, their lifestyle is psychologically comparable to the routine tasks required for agriculture, so that an intelligent gibbon might make a better farmer than a baboon. Hunter-gatherer peoples such as the Mbuti enjoy hunting and gathering but lack the necessary temperament for farming.

This problem is not especially acute for slash-and-burn agriculture, as practiced traditionally in places like New Guinea or the Amazon, and which takes at most only a few hours a day. But civilization depends on growing large amounts of food per hectare, which requires an intensive work schedule day after day, month after month, year after year. The same applies to the routine work of factory and office. Therefore, maintaining this temperament in civilization is of the utmost importance.

The behavioral and physiological complex associated with food shortage is even more vital to civilization when we recognize that it has another effect—making people more capable of impersonal loyalties. Societies with behavioral signs of food shortage, such as nuclear monogamous families, late breeding and control of children, also tend to have larger political units

Such attitudes have no direct equivalent in monkeys or apes, but we can understand them by inference. For example, a baboon's loyalties and ties are to members of its band, and so too are the ties that bind hunter-gatherer societies. In other words, they are *personal*.

A gibbon, by contrast, is attached only to its mate and offspring, and beyond that to a piece of land—its territory. This latter attachment, which baboons do not have, is by nature *impersonal*. Large-scale societies will be more stable if people can be loyal to institutions and to people they do not know.

As loyalties shift from the personal to the impersonal, there is an increasing potential for political organization that does not rely on personal ties.

At the most personal level, as in a hunter-gatherer band, loyalties are felt to friends and relatives who are seen every day. A slightly more impersonal loyalty might be to a tribal chief or feudal lord, who may be personally known but not nearly as well. Someone with a still more impersonal orientation might accept the authority of a king or emperor—an identifiable individual, but one who might rarely or never be seen. And at the most impersonal, there can be loyalty to the laws and institutions of a state rather than any individual as such. This is why people in a democratic society accept the authority of a leader even when they voted for the opposition. It also allows them to form efficient and impersonal bureaucracies.

The same concept of impersonality also explains why societies with behavioral indications of food shortage tend to have market economies. Markets and money are impersonal ways of distributing goods, as compared to more personal motivations such as friendship, reciprocity or fear, which tend to form the basis of trade and distribution in smaller scale societies.

Hard-working farmers, larger states and markets are benefits of the temperament that comes from the food-shortage mechanism. When it comes to competition between different groups, civilized societies have huge advantages in the struggle for survival. With better-organized states and dense, hardworking populations, they tend to sweep aside smaller scale societies and hunter gathers with relative ease. But to achieve all this people have to behave more like gibbons and less like baboons, which for humans is very unnatural behavior. Societies which come out on top are those in which the physiological complex associated with food shortage has been activated.

But how is this possible? Civilized societies are not universally more short of food than simpler ones. In fact, as we will see, the one society in world history with the most extreme form of "food short" behavior—nineteenth-century Europe—was actually quite well off.

What makes it possible is that hunger is not the only way to activate this complex. Suppose that thousands of years ago a people had found that scattering a particular kind of seed in turned earth brought about a better wild crop some months later. Because such work is boring they did this only fitfully.

But imagine that one local band adopted a custom that, quite inadvertently, mimicked the physiological effects of food shortage *without actually limiting food*. This in turn caused the women to spend more time digging earth and scattering seed. As a result, the band gained numbers and prestige, either colonizing nearby areas or causing neighbors to adopt similar practices. As the people began selecting the best seeds for planting and thus improving the varieties, so the advantage of hard work increased. Groups within the wider population where customs made men more likely to help did better still. As the process continued, certain groups became able to work together more effectively, adding political to economic strength, and so civilization arose.

Of course, the people would not understand *why* this worked. Most likely, they simply believed that their god required them to act in a certain way and would bless them if they did and blight them if they did not. It is an attitude commonly expressed in the Book of Genesis and throughout the Hebrew bible. And, of course, they were right. Acting in certain ways did help the Israelites to prosper, as we will see, though in this case any direct divine intervention would be greatly aided by the fact that these customs *worked*.

The rise of farming and civilization was made possible by physical technologies, such as food plants, domesticated animals, irrigation systems, roads, markets, metal-working and writing. But equally vital was the culture that created the civilized temperament, and which became steadily more powerful and effective over the next ten thousand years.

For convenience, biohistory refers to the array of traits that makes up this civilized temperament by the shorthand label "C," because people with high C have a temperament better suited to the needs of civilization. A person or group of people can have high C or low C, depending on environmental pressures and the strength of the cultural technology.

In short, high C is a physiological system that makes people harder working, better at trade, and more supportive of strong states. It also causes them to delay breeding and form nuclear monogamous families.

C and testosterone

Before we can understand how individuals and thus societies achieve a higher level of C without food shortage, it is important to understand how the C system works. In the terminology of biohistory, anything that raises C is a "C-promoter." A C-promoter can be identified as *anything that reduces testosterone*. This does not mean that high C societies always have low testosterone, because high C parents treat their children in ways that can have quite different effects (as we will see later). But in terms of their effect on adults, this is the easiest way to identify a C-promoter.

We have seen from human and animal studies that mild hunger reduces testosterone, which makes hunger a C-promoter. People with low testosterone tend to be less sexually active, less aggressive, more passive in face-to-face encounters, and have more stable sexual partnerships. They take a greater interest in their children and are more likely to control their behavior. They are also less outgoing, and tend to socialize less.

High testosterone can be an advantage in a hunter-gatherer society, in which men may work for as little as five hours per day, compared with (say) 8.2 in Britain in 1800.[30] In such a society the most successful men are usually the most aggressive and dominant, good at hunting and war, and successful at attracting fertile women.

In civilized societies the advantages are largely reversed. Men with lower testosterone tend to excel at jobs which require hard work and delayed pay-offs. One example is farming, another is engineering, and yet another is business. Success in all such fields is associated with lower levels of testosterone and the high C temperament that goes with it. Such people tend to command higher salaries. The only occupations that favor high-testosterone men are sports, entertainment and to some extent the military. As discussed in the last chapter, high-testosterone men are more likely to be unemployed.

This is contrary to the popular idea that dominant and high testosterone men are likely to be more successful, but it is what the evidence shows.

How culture promotes C

Understanding this, we can now look for C-promoters in human societies, which means cultural customs that reduce testosterone and thus activate the complex of behaviors associated with chronic food shortage. C-promoter

Fasting

The most obvious of these is fasting, which is of course an artificially induced hunger. Fasting is prescribed by a number of religions including Judaism, Catholicism and Orthodox Christianity. Christians have traditionally been expected to observe the Lenten fast in the weeks leading up to Easter. Buddhist monks and nuns are expected to not eat after midday, and Hindus, Jains and Bahais all practice various forms of fasting.[31]The Muslim fast of Ramadan is particularly rigorous, and believers are expected not to eat or drink at all during daylight for a whole month. Islam teaches that fasting promotes patience, self-discipline and self-control.[32]

Choosing low calorie food has a similar effect. The Lenten fast of some Christian churches mainly involves avoiding certain kinds of food, such as meat. The same applies to complete or partial vegetarianism, as practiced by many Hindus, Buddhists and Christian groups such as Seventh Day Adventists. The Jains in particular are rigorously vegetarian, and it is no coincidence that they are commercially successful. Though making up less than 0.5% of the population of India, they pay 24% of the country's income tax.[33]

Fasting is a powerful C-promoter, but alone it is not enough. Taken to an extreme it can undermine health. Also, eating is a somewhat private activity and not easily subject to social controls. And finally, it almost certainly is not powerful enough to create the ultra-high levels of C required by the most successful human societies.

Restricted sex

Fortunately for the development of human civilization, there is another and far more effective C-promoter that has none of these drawbacks. This is *restriction of sexual behavior*. We already know that lower testosterone reduces sex drive. A less obvious finding, but one supported by scientific evidence, is that restricting sexual activity reduces testosterone.[34] To take one example, a study of couples showed that testosterone levels rose in couples who had sex over a given period, while those who abstained showed a decrease.[35] In humans and animals, sexual arousal increases not only testosterone but the level of luteinizing hormone, another hormone associated with hunger.[36]

Sexual abstinence is a C-promoter in that it reduces testosterone in adults. Like fasting, it is a practice prescribed by religion. Restriction of sexual activity has been a feature of all known civilizations, and loosening standards tend to be followed by a decline—as in the case of Rome in the period between the middle Republic and the Empire (see chapter twelve). Warrior cultures, such as those of Afghanistan, support C by strictly controlling the sexuality of women, while maximizing the testosterone of men by allowing them greater freedom. But more generally, and especially for groups focusing on commerce such as Jains and Jews, both sexes have far less sexual activity than they would like.[37]

The age at which sex is first experienced seems to have an especially great long-term impact on testosterone. One study found that rats that were sexually active shortly after puberty had double the testosterone of those without such experience, even eight months later and without exposure to females in the meantime.[38] Such an experiment has clearly not been done for humans, but the Kinsey Report indicates that young men who deliberately curbed their sexual behavior as teenagers (as indicated by the level of nocturnal emissions) were more likely to be successful in tertiary education and professional occupations.[39] And we have seen that occupational success is associated with lower testosterone.

A common explanation for the control of women's sexuality is to ensure paternity, and this undoubtedly provides psychological support for the practice. But it does not explain the extreme lengths taken by so many societies, nor why frequent sexual liaisons can be found in some highly patriarchal societies such as the Yanomamo of the Amazon (as we will see in chapter 5). A better explanation is that limiting sexual activity, especially in young people, raises C and thus helps the society to be more successful.

As mentioned in chapter one, a number of studies have found a link between sexual restriction and civilization. It is significantly correlated with larger political units, the capacity for hard work and market economies, and also with the "biological" aspects of C such as monogamous nuclear families, late marriage and control of children.[40] Previously considered as an indication of C, we now recognize it as also being a powerful promoter of C. Its value as such depends on reducing sexual activity well below the level that is "natural" and congenial, which makes it very hard to maintain. Only the strongest religious and social sanctions can achieve this, and even these commonly break down in wealthy urban societies where plentiful high-calorie foods reduce C.

Control of children

A third type of C-promoter is the *control of children*. Aside from epigenetic inheritance or influences within the womb, this is the main way in which C is passed from one generation to the next. Attentive and disciplined child-rearing affects the physiological and epigenetic development of children, and therefore their temperaments and lifelong behavior. The pheromones of parents who are in constant contact with their offspring may also play a role.

Parents with high C tend to be hard working and disciplined, a temperament which also makes them more likely to discipline their children. But the wider culture can help reinforce the C of each generation by pressuring low-C parents to conform and exert control over their children through community pressures, religion, schooling and even legal sanction. Thus, a high-C society can raise the C in even its lower-C individuals and groups.

As with sexual restraint, a society experiencing a long-term decline in C will tend to be more permissive with children. Parents who are low in C and lack the reinforcing pressure of a high-C society allow their children greater freedoms and less discipline. (Western societies are presently in such a decline, and have been for many decades. In chapter sixteen we will examine in detail the complex temperamental shifts that have brought this about.)

There have been suggestions in recent years that genetic change could account for the increase in C-type behaviors in Europe between about 1200 and 1850, but the rapid collapse in C since then indicates that this is primarily an epigenetic rather than a genetic change.[41]

The effect feedback cycle—how C promotes C

C-promoters are to some extent self-sustaining. High C parents tend to discipline their children, which transfers C to the next generation. People who restrict their sexual behavior, whether through personal choice or cultural pressure, have lower testosterone. This reduces sex drive, which makes it easier to restrict sexual behavior. That is a very simple form of feedback where a C-promoter (sexual restraint) raises C, which makes the C-promoter easier to sustain.

The same applies to food. Eating less than one would like, for religious or cultural reasons, increases C. This in turn makes disciplined eating easier. This observation is called the *effect feedback cycle*, as illustrated in Fig. 3.1.

Fig. 3.1 The effect feedback cycle

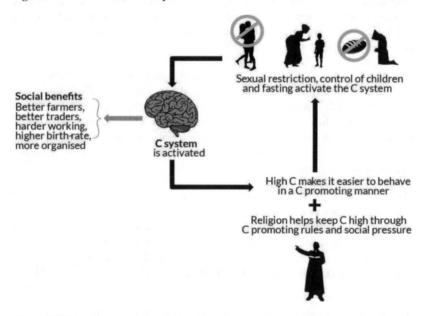

Social benefits
Better farmers, better traders, harder working, higher birth-rate, more organised

C system is activated

Sexual restriction, control of children and fasting activate the C system

High C makes it easier to behave in a C promoting manner

Religion helps keep C high through C promoting rules and social pressure

But C-promoters are not completely self-sustaining. Higher C people may find it easier to restrain their sexual activity or eating than those with lower C, but that does not make it easy. To maintain C they will still need to have less sex and eat less food than they would otherwise like. High C can only be maintained by considerable cultural pressure.

The reason for this is that humans are naturally a low-C species. As noted in chapter one, our genetic set-point for hunger-behavior is more like that of baboons than of gibbons. To maintain C at unnaturally high levels is profoundly difficult, which is why it has taken human societies ten thousand years to learn how to do it.

We have seen that C-promoters such as eating less and having less sex are also effects of lower C, and vice versa. But it appears that the same applies to any other form of behavior linked to C. This means that any form of behavior promoted by higher C can also act as a C-promoter.

For example, people with higher C find it easier to work hard, whether on a farm, in a factory, as a student or running a business. But a person who engages in such work *beyond the level set by temperament* supports their own C by doing so. Hard work is thus in itself a C-promoter. The cultural and economic pressures for people to work harder than they would like are, of course, enormous. Hard work is not enough to support C by itself, but it is a contributing factor.

The Role of Religion

People with high C tend to be disciplined and orderly, so any code of behavior can act as a C-promoter, such as codes of etiquette and manners, standards of dress, getting up at a set hour, and so forth. But the key C-promoters are those found in religion. Fasting and restraints on sexual behavior are key elements of all major religions systems, including Buddhism, Hinduism, Islam, Judaism and Christianity.

Other religious practices have a similar effect. One of the most effective is Sabbatarianism, the practice of working for six days of the week but taking the seventh off. Others serving as C-promoters include church services and processions, set prayers and scripture study. All require behavior and even thinking to be disciplined and controlled, often to a very precise degree. Religious C-promoters are all the more effective because social disapproval is backed by the sense of a powerful and all-seeing deity. It also helps that higher C increases religious feeling and the sense of divine presence, something obvious to anyone who has attended a church service following a fast.

The elaborate structure of Jewish ritual law, which controls every detail of its followers' lives, is a powerful C-promoter. In three millennia of history, Jews have often been on the margins of society, to the extent that their very existence has been in jeopardy. Yet each generation has encouraged the best and brightest of young men to pursue rabbinical studies, supported by wealthy fathers-in-law and the community. At one level it might seem to make better sense to divert such resources into making a living, but not if we understand the real value of Jewish law. By maintaining ultra-high C, which is especially valuable for success in commerce, Jewish ritual law becomes one key reason for the survival and success of the Jewish people. This is why Jews who abandon the law normally lose their identity within a few generations.

This brings up the more general topic of the value of religion. Scientists such as Richard Dawkins preach that religion is the outdated superstition of another age, without value or purpose in civilized times. But a better understanding of the science indicates that religion is the essential driving force behind human civilization, which would not be possible without it. Religious teachings not only help people to cooperate better in large groups, they create the very temperament that is needed if civilization is to flourish. That is why the decline of religion is a danger sign for any society.

How C declines

Humans are naturally a low C species, so maintaining high C means that people must act in ways very different from what they would like. Only the most intensive social and religious pressures can keep such behavior in place. This means that any factor that reduces C makes such behavior even more difficult to maintain, which means C-promoters such as sexual restraint are loosened, which lowers C still further, which causes still more loosening of C-promoters, and so the process continues until the civilization collapses.

It is as if two groups of men were equally balanced in a tug of war, straining mightily to keep the rope in place. Add a single extra man to one side, even a not very strong one, and the other side starts to give way—slowly at first, but then faster and faster as more of them slip and fall over, until the gradual slide becomes a rout.

The one force most likely to upset this balance is wealth. Plentiful food and material comfort have the opposite effect to C-promoters and may be termed 'C-demoters'. An initial reduction in C by affluence not only makes high C behavior more difficult but lessens the religious sentiment that helps support it. There is only so far this process can go in a single generation because the basic level of C is set by parental control, but over a series of generations the process is inexorable. As C declines and traditional C-promoters collapse, ever more powerful C-demoters are sought, such as high calorie foods leading to an obesity epidemic, and addictive drugs such as heroin and cocaine.

To take just one example of how calorie intake can affect C, it has been found that adolescents who drink more than five cans of soft drink a week are more likely to carry a weapon and act violently with peers, family members and dates. Other studies have linked soft-drink consumption with violence and mood problems in adolescents, and with aggression, withdrawal and attention problems in young children and teenagers[42] All these behaviors indicate lower C.

Civilization has developed over the past four thousand years through the growth of religious traditions with ever more powerful C-promoters. The weakness of this process is that people are required to act in ways that are very different from those congenial to our genetically set temperament of low C. Civilizations fall, at least in part, because wealth upsets the balance, pushing down C and thus undermining the C-promoting behavior that supports C and thus the civilization itself.

I say "in part" because there is another factor which supports C, which accounts for the very rapid rise of C in the West up to the nineteenth century, and whose loss accounts for the even faster decline of C since then. This has to do with stress, and is the subject of the next chapter.

Testing

Limiting the sexual activity of human volunteers for a few weeks should produce physiological changes associated with C, and also changes in the attitudes and behaviors associated with high C, such as religious sentiment, the capacity for work and self-discipline, and reduced sociability.

CHAPTER FOUR

AGGRESSION

Stranger, go tell the Spartans that here we lie, obedient to their commands.
—Monument at Thermopylae

In the previous chapter we saw how human societies have developed cultural methods for increasing the level of C, giving people a temperament that makes it more likely that they will work hard, run successful businesses, have wider political loyalties and prosper within civilization. The methods used include fasting, rituals and above all limits on sexual behavior, and are commonly associated with religion.

But successful civilizations are built upon more than just hard work and trade; more than invention, productivity and discipline; more than political loyalties and nuclear families. And naturally it follows that there is more to civilization than a high level of C. A successful civilization needs to be able to overcome its enemies. It needs some level of warlike aggression.

We saw in chapter two that C-promoters can lower testosterone and thus reduce the level of aggression, but high C societies are not especially peaceful. The British of the Victorian era had exceptionally high C, as indicated by the levels of industriousness, discipline, intellectual achievement, invention and productivity. This was supported by the Victorian moral code, which is a byword even today for its control of children and sexual restraint. But the courage of British soldiers at this time was extraordinary by any standards, as indicated by accounts of the Crimean War in the 1850s.

This war is perhaps best known for the incompetence of the British generals, but it is also an astonishing account of raw human courage and vitality. One outstanding engagement was the charge of the Heavy Brigade at the battle of Balaclava, in which five hundred British cavalry charged uphill into three to four thousand Russian cavalry, so outnumbered that they were completely engulfed in the enemy lines. There was no space for pistols or carbines. The men hacked at each other with swords or, when these broke, fought with their bare hands, yelling and cursing with rage that could be clearly heard on the heights above. The enemy broke, and then fled. The Russians, brave soldiers as they were, were shocked and in awe at such an impossible defeat.

In another engagement in the same war, 550 Highland infantry, deployed in two long ranks, repelled a charge by a Russian cavalry force that outnumbered them eight times over—a stand which led to the coining of the expression, "the thin red line." At one stage the Highlanders wanted to leap out and charge the enemy, and were only with great difficulty kept in their places ("Damn that eagerness!" their colonel shouted at them). Most famous of all was the charge of the Light Brigade, in which seven hundred British cavalry made a suicidal charge and were cut to pieces by Russian artillery. The Russian commander, when interviewing prisoners, assumed they had been drinking. He could think of no other explanation for the blind courage displayed.[43]

Warlike courage is also typical of other high C peoples such as the Romans of the third and second centuries BC. In fact, a people's level of aggression seems to be relatively independent of their level of C. Consider two societies with relatively low C but at opposite extremes in terms of their level of aggression.

The first is the Yanomamo, already considered in chapter two as an example of a low C society with no political organization beyond the village level and practicing only part-time shifting agriculture. They are also ferociously aggressive, feared by their neighbors and engaging in incessant warfare amongst themselves.[44]

A relatively moderate expression of aggression comes in the form of mock warfare where members of opposing tribes, or groups within a tribe, strike each other in the chest. Competitors often cough up blood for days after and fatalities are not uncommon. Beyond this, club fights break out at the slightest provocation and can quickly engulf an entire village. There is also outright war. Raiding between villages is constant, usually justified by some past grudge. Even if there has been no actual offense to justify revenge, they may attack on the belief that the other village has been using sorcery against them. Fear and suspicion of outsiders is endemic within Yanomamo society and not without some cause—it has been estimated that 25% of all deaths among adult males are due to violence.

Success in warfare gives great benefits. Men who have killed produce, on average, three times as many children as those who have not, due to their success in seizing and keeping multiple wives.[45] In order to succeed, a Yanomamo man needs to be extremely ferocious.

Consider by contrast the Tchambuli, a New Guinea people at a similar state of economic and political development. As described in Margaret Mead's colorful account, males in this society could hardly be more unlike their Yanomamo counterparts:

> All that remains to the individual Tchambuli man, with his delicately arranged curls, his handsome public covering of a flying-fox skin highly ornamented with shells, his mincing step and self-conscious mien, is the sense of himself as an actor, playing a series of charming parts—this and his relationship to the women. His relations to all other males are delicate and difficult, as he sits down a little lightly even in his own clansmen's house, and is so nervous and sensitive that he will barely eat in the houses of other clans ....[46]

Rather than winning females by ferocity and fear, Tchambuli males compete with each other on the basis of charm and looks! They are so lacking in aggression they are forced to maintain their headhunting traditions by *purchasing* captives from neighboring tribes. Later studies suggest that this overstated the case, that the Tchambuli did make war at times and that many men beat their wives.[47] But the Tchambuli do seem to have been considerably less patriarchal *and* less warlike than their neighbors, as indicated by the fact that in the immediate pre-contact period they had been driven from their fertile lakeside home.

It is significant that the Tchambuli lived in affluence (by pre-industrial standards) because historically, the most warlike peoples have tended to come from lands with harsh climate and terrain. These include the Elamites and Persians from the mountains or Iran, the Aramaeans and Arabs from the desert, and the Mongol horsemen from the vast steppes of central Asia. Again and again they have swept down to raid and rule the more peaceable peoples of the settled lands, only to lose their warlike aggression and be raided and ruled in turn by a fresh wave of "barbarians."

It is important to recognize that the ability to make war successfully, whether as a Yanomamo warrior or a British soldier in the Crimea, requires not just aggression but the ability to fight *as a group*. Individual violence, simply slashing out at anyone nearby, would be quite ineffective as a way of defeating external enemies.

We need to understand what makes some peoples more warlike than others, why this temperament tends to be lost when people become civilized, and how the British of the Victorian era managed to maintain their capacity for war in a prosperous and urban society.

Aside from C, biohistory posits another physiological system which affects temperamental traits such as competition and aggression. These traits are grouped under the umbrella of *vigor*, and are denoted by the label, "V." Societies with low C can be high V and ferocious like the Yanomamo, or low V and more peaceful like the Tchambuli. The term *vigor* is used because V includes not only aggression but also other assertive characteristics such as pioneering spirit, and the urge to expand and migrate. But aggression is connected to these characteristics (through expansion and migration with conquest), and is therefore the main focus when discussing a society's level of V.

Some societies have levels of V and C that are inverse to each other—one is high, the other low. And then there are societies where both C and V are high. These are economically advanced civilizations with vigorous and aggressive characters. One example is Victorian Britain, as indicated earlier, but there are many others.

Where does V come from? We have established that C is raised in adults by any influence that lowers testosterone, and is transmitted to children mainly by parental control. What is the basis of V?

Baboons

Once again, as with C, animal models can help to make this clear, and the model we will use for high V is baboons. Baboons are perhaps the most ferociously aggressive of all primates and thus a good example of high V.

Consider their reaction to leopards:

> [If] the baboons are able to isolate a leopard in a bush, tree, or aardvark hole, they immediately surround it, screaming, alarm-calling, and lunging at it, seemingly without fear. Although male baboons, with their size and enormous canines, are much better equipped than females to fight a leopard, the mass mobbing involves baboons of every age and sex. Juveniles, adult females, even mothers with young infants join to form a huge, hostile mob that tries to corner the leopard. The attack continues even after some baboons have received slashes on their arms, legs, and face that open up huge wounds.[48]

Leopards are not uncommonly killed by such attacks. Baboon males also fight savagely amongst themselves when contesting for dominance in the troop. But when the hierarchy is stable, as is most commonly the case, they are able to act very effectively in concert. Troops even move in a highly organized fashion. They set off in a coordinated manner with males—especially the alpha male—somewhat more likely to take the initiative,[49] and with the most vulnerable troop members protected in the center.[50] When the troop is faced with danger the dominant males race forward to face it, provided they have reason to believe some vulnerable infants are theirs. Just like the human societies most successful in war, baboons are both highly aggressive *and* well-organized.

Baboons are also low C, which helps to make them bolder, but this does not account for their ferocity nor their organization. Domesticated animals are also low C in that they are highly sexed, breed fast and provide little care of young. But they are also strikingly unaggressive, which is vital if they are to survive the intense crowding of the farmyard.

Baboons are not only aggressive but intolerant of crowding. Early last century a colony of mainly male hamadryas baboons was established at Whipsnade Zoo in England. They became intensely stressed and most were killed in vicious conflicts, especially the vulnerable females and young. Status differences were extreme and yet the status hierarchy was unstable, all behaviors which we will find associated with stress in human societies.[51]

Thus, while domesticated animals are low C and low V, baboon are low C and high V.

The Biological Function of V

In chapter two we saw that C is increased in animals by mild, chronic food shortage, and it adapts behavior to environments where food is chronically short. V, on the other hand, is an adaptation to dangerous environments where food supplies are highly *variable*.

Consider the savannah habitat where most baboons live. As discussed in chapter two, food for baboons is plentiful for the most part but in times of drought can vanish almost completely, leaving them to starve. When food is unavailable locally the only way to survive is to migrate to areas where some may be available. And what this requires is aggressive, tightly organized bands which are energetic enough to make the journey. In this sense, high V behavior is completely adaptive.

The other danger baboons face is from predators. Most monkeys live in trees or at least in forested areas where they can easily escape from predators. But baboons are ground dwellers who live with the constant threat of attack by lions and leopards. The level of resulting anxiety is illustrated by this account of baboons crossing a dangerous water channel:

> Water crossings … are fraught with anxiety. Long before they enter the water, the baboons sit at the island's edge, nervously grunting and looking out towards the island they hope to reach. Any movement on the water's surface elicits a chorus of alarm calls and brief flight. Once they seem satisfied that the coast is clear, adults begin to cross. Reluctantly, the juveniles follow, some grunting nervously, others moaning or screaming, and others running to leap on their mothers' backs, anxious to get a ride … The whole spectacle is chaotic and amusing to the human observer but deeply distressing for the baboons, who are out of their element and vulnerable to any predator that lurks in the water or along the too-well-travelled path.[52]

Actual attacks are even more traumatic, and analysis of feces shows that female baboons experience a dramatic rise in stress hormones after a close relative is taken.[53]

The aggressive and well-organized high V character of baboons not only makes migration feasible in times of famine but is an effective way to deal with predators. What both these dangers—famine and predators—have in common is that both are highly *stressful*. And this will give us a clue as to the source of V.

Intermittent stress increases V

Stress can best be seen as a response that helps animals cope physically with danger—the so-called "fight or flight" mechanism. When faced with a challenge, such as a hungry lion or a rival member of our own species, stress hormones including adrenaline, noradrenaline and cortisol are released into the bloodstream. These work to shut down bodily systems not related to fighting or running away, diverting resources into muscles and other functions needed to cope with immediate danger.[54] Normally, this reaction shuts down quickly when the danger passes, but in some cases stress can become chronic, such as when an individual cannot escape from being harassed by more powerful rivals.[55] In this case the response system becomes weakened, resulting in psychiatric disorders.[56] Dopamine is reduced, leaving people with low motivation and unable to experience enjoyment. Depleted noradrenaline cells can lead to a lack of arousal and enthusiasm.[57] Depressed people commonly combine low levels of noradrenaline and dopamine with chronically high cortisol.[58]

However, short-term stresses—such as those experienced by baboons— have a very different effect. They *increase* the body's ability to manufacture dopamine, noradrenaline and adrenaline, which act very quickly in readying it to deal with danger.[59] A person with an adrenaline/dopamine response to stress is more likely to see it as a challenge. Short-term stresses also bring about a more effective, less harmful cortisol response.[60] Cortisol ramps up when needed but dies down quickly when the immediate challenge has passed, thus avoiding the damaging effects of long-term stress. It is these short-term stresses which give the body time to recover that enhance its capacity to cope with future stress.[61]

Furthermore, more efficient cortisol production helps to moderate aggression, offsetting the negative effects of high testosterone. In other words, the individual is dominant, assertive, competitive and *capable* of aggression, but without the uncontrolled, antisocial aggression that comes from chronic high testosterone. . Such controlled aggression is what allows high V individuals to cooperate with each other and fight off external enemies. Environmental stresses such as famine and predators thus produce *exactly* the right behavior to cope with famine and predation.

Authority, stress and V

Now we must consider how generations transmit V to one another. There is an obvious advantage to this, in that famine-prone populations need to be able to migrate effectively and fight off predators, even when some individuals may never have experienced either.

One of the patterns seen in the study of primates with different levels of C—such as baboons and gibbons—is the difference in how offspring are cared for. In general, the pattern is that higher-C species (e.g. gibbons, adapted for food-restricted environments) produce fewer offspring and give them more attentive care, whereas lower-C species (e.g. baboons in environments with plentiful food, with periodic scarcity, and danger from predation) breed much more rapidly and give much less attentive parental care. Similar behavior can be observed in species as diverse as rats and humans, though in humans the main C-promoter for children is parental control (see chapter two).

That is the general pattern, but not the whole picture. Baboons, for instance, do not neglect their infant offspring. Baboon babies are given a great deal of close care and attention by adults during the first year or so of life, particularly from their mothers.[62] But it doesn't last long. Maternal care continues, gradually declining in intensity, until the infant is weaned at about a year old. At this point, maternal care ceases, the mother goes into estrus, ready to breed again, and the infant is cast off and left to fend for itself.[63]

This neglect imposes stress on the young baboon but, as noted earlier, stresses can have different and even opposite effects. They can lead either to chronic anxiety or to a toughened and enhanced stress response. When it comes to early life, what matters most is the form of stress and the *age* at which it is experienced. Neglect or punishment in infancy tends to result in lifelong anxiety, including depression and withdrawal or extreme, uncontrolled aggression. But intermittent stress *after* weaning, which includes any experience of powerful authority, seems to produce the kind of toughened and enhanced stress response seen in baboons.[64]

The long-term effects of early stress seem to depend on whether the child is old enough to understand the initial stress and can *do something to avoid it*. Infants are not capable of either understanding or avoidance. A punished or neglected infant can do nothing to avoid or defuse the source of stress, and therefore the ability to cope with stress in later life is severely compromised. This problem was discovered quite inadvertently when rhesus monkeys were raised without mothers so as to produce healthy animals for research. The project failed, as the animals were so maladjusted as to be useless for experimental purposes. They were unable to relate to others, incapable of normal sexual behavior and liable to self-harm.[65]

But the situation changes after weaning. Among baboons, stresses on juveniles stem from attacks by other juveniles or higher status adults. In humans, the most common factor increasing V in late childhood is adult authority, either in the form of punishment or control. This is not to say that punishment and control have exactly the same effect. Although both increase V, control based on psychological methods has less impact on V than control based on severe punishment.

But for either to be a V-promoter, the juvenile or older child must have some ability to avoid the stress either by running away or by ceasing the behavior that brings it about. If they cannot avoid it, such as when chronically punished or abused for no consistent reason, then the toughened (high V) stress response is not formed.

The Mundugumor tribe of Papua New Guinea, studied by the anthropologist Margaret Mead in the 1930s,[66] provides an example of what can go wrong in a society when chronic stress is inflicted on infants. Their treatment of even the youngest infants was rough and neglectful. Infants were kept in a basket hung from a peg and had little contact with their mothers. Older children were subjected to severe and inconsistent punishment. The death of a child was regarded as an annoying mishap rather than a tragedy.

Mundugumor society was riven by hostility and mistrust at every level— between parents and children, husbands and wives, brothers and neighbors. Fear was a constant, ranging from fear of other people to fear of drowning. The tribe was ferociously aggressive and feared by its neighbors. In short, the Mundugumor showed all the features associated with early and chronic stress including fearfulness, extreme aggression, and poor social adjustment.

This is very different from the way young baboons are treated, which combines intense care of infants with harsh treatment after weaning, and has very different effects. The Mundugumor as a people were declining in numbers, while baboon behavior tends to maximize population growth. The vulnerable infants are guarded, but by weaning as soon as possible the mother can go back into estrus and produce another baby. Fast breeding is, of course, a key requirement for any population in a dangerous environment. We will see a high birth rate as one outstanding characteristic of high V human societies.

Patriarchy

There is one other kind of stress that is very effective at promoting V, perhaps the most effective of all. This is when an individual experiences a high level of anxiety in infancy, *not* as a result of neglect or abuse, and then goes on to experience much lower anxiety as an adult.[67] This can happen in a number of ways, such as when the level of anxiety is dropping rapidly in the population, but the most common circumstance in human societies is when an infant male is close to a very anxious mother, and then grows up to be a much less anxious man. In other words, women must be made more anxious than men. This is achieved by making them subordinate to men. In other words, V (in males) is promoted by patriarchy.

This is the case in all the high V societies described so far. Baboon males are strongly dominant over females, helped by the fact that they are around twice the size. That Victorian Britain was a patriarchal society hardly needs mentioning. Yanomamo men are probably even more dominant. A wife who is slow to prepare her husband's dinner may be scolded or beaten by hand or with a piece of firewood. More severe punishments for women can include being cut with a machete, shot with an arrow in the leg or buttock, burned with a glowing stick, or even killed (the latter usually for infidelity or even just the suspicion of it). At times, Yanomamo men beat their wives for no apparent reason, other than to "keep them on their toes."[68]

The Tchambuli, by contrast, are one of the few societies in which women (at least in Mead's opinion) hold effective power. Men depend on women for their food, and a man must gain the approval of his wife before making a purchase. This is particularly evident in attitudes towards property, which men must beg from women in return for "languishing looks and soft words."[69]

The role of patriarchy is that it not only creates a high level of anxiety in infancy, but leads to much lower anxiety in adult males. Once a person becomes an adult, the high level of V they have built up through early stress will be reinforced and heightened if that person has a high status in society. High-status people experience less chronic stress, and thus have a more efficient cortisol response. As with C, V is sustained and enhanced by an effect feedback cycle, whereby behavior caused by high V tends to promote V. For instance, dominance and competitiveness are increased by V, and also cause intermittent stresses which enhance V.

V-promoters

V is a product, then, of a specific set of influences which may be termed "V-promoters." But while C-promoters have the link to food shortage and reduced testosterone in common, V-promoters are all related to authority and stress. These include an anxious but affectionate mother, intermittent stresses and exposure to adult authority in late childhood, and ongoing stresses but relatively high status as an adult.

V-promoters in human culture include patriarchy, support of physical punishment through such principles as "spare the rod and spoil the child," reinforcement of parental authority by admonitions to obey parents, and fasting. But the kind of fasting that promotes V is not regular, short-term fasting of the kind that maximizes C. It is more severe, annual fasting that mimics the effect of an occasional famine. Examples are the Lenten fast of traditional Christianity, or (most effective of all) the Muslim fast during Ramadan.

One final V-promoter in human societies is control of women's sexual behavior, possibly because increased C makes women less likely to neglect infants even when stressed by their low social status. This is clear from cross-cultural studies.[70]

The result is a temperament which makes people both aggressive and well-organized—a character built for war. Not all of these characteristics are necessary for a high level of V. The Yanomamo are generally indulgent with children of all ages, though they may inflict harsh and ill-tempered punishment at times. Victorian parents, by contrast, were patriarchal and punitive enough but remarkably lacking in indulgence of infants, so it is no coincidence (as will be discussed later) that their level of V was to plummet in the century ahead. But in general, the stronger the V-promoters the higher the level of V.

The benefits of V in human society

Although the individual has framed much of the discussion so far, V and C are best understood in terms of social groups. Personality is a complex matter with a multitude of genetic and environmental influences. Therefore, V is best considered as a factor that distinguishes one social group from another, and also one that describes how societies change with time.

At the most basic biological level, V is a set of behavioral and temperamental characteristics that helps animals survive and thrive in dangerous, unpredictable environments. It includes traits such as a high birth rate, aggression, competitiveness, morale, the ability to cope effectively with challenges, and the organization and boldness required for successful migration.

In human societies, the most obvious indicator of V is a society's attitude to war. In general, higher-V societies have a strong advantage in any conflict with lower-V rivals. Among human societies, the highest V peoples are the warlike tribes which have so often invaded the settled lands from mountain, desert and steppe. These are the environments which not only increase V by periodic famines, but create further stresses when the resulting warlike tribes make war with each other. High V not only makes them aggressive and well-organized, it also maximizes population growth. By combining the intense care of infants with abrupt weaning to allow women to become pregnant or effectively nurse the next child, high V compensates for the extreme dangers of the environment that gives rise to it. Thus, these aggressive, warlike and fast growing populations are prolific and effective migrants.

One culture almost wholly devoted to heightening its V was ancient Sparta. In the account given by Plutarch in his biography of Lycurgus, the Spartan lawgiver, he describes the military training of boys. They were given constant, vigorous exercise, including wrestling and fighting. They were allowed only a single tunic and cloak in the coldest weather. They were left short of food, so had to steal or starve. They were trained to endure pain, such as in a public ceremony where boys ran a gauntlet of flogging to steal cheeses from an altar.

These are all severe stresses but they are, by nature, intermittent. Wrestling and fighting are occasional activities. Extreme cold would not be experienced all year round. Hunger would only be occasional, relieved when they were able to steal or otherwise get hold of food. Flogging at the altar happened only once a year. Spartan boys were subject to the strictest and most rigorous forms of adult authority, but as adults they had relatively high status compared to their helot subjects, who could be murdered at a whim. Put together, what this means is that Spartan warriors had much higher V than their rivals, whose armies comprised part-time soldiers.[71]

The control and discipline of children in Victorian Britain also produced a high level of V. From strict nannies and governesses in the home to the rigors of a typical English boarding school, the ruling classes subjected their children to a powerful V-promoting regime:

> Almost unconsciously the public school boy [a British "public school" is a private boarding school, not a public school in the American sense] absorbed a complete code of behavior which would enable him to do the "right thing" in any situation. It involved obedience to superiors, the acceptance of a position in the hierarchy, team spirit, and loyalty ... It also involved the traditional British phlegm, reserve, understatement,

unflappability, the stiff upper lip, a result of the inculcation of modesty in victory and defeat, the all-male society in which emotion was sissy, the encouragement of restraint in the exercise of power.[72]

These were the boys who grew up to form the officer class of the British Army and Navy in the Victorian era. From platoon commanders to generals, from midshipmen to admirals, every British officer had gone through the public school system, or some equivalent of it (naval colleges were run on the same disciplinary model, as was the military academy of the East India Company).

Military effectiveness can be regarded as controlled, marshaled aggression. This is a central characteristic of V. Military drill—including weapons training, tactical exercises, and parade-ground drill—increases V. Many observers, including new recruits, wonder at the purpose of marching in step and presenting arms. How does incessant drill make better soldiers, especially since marching in step is not the way modern armies fight? But experience has shown it to be an effective way of building morale and the general effectiveness of soldiers. Robert Graves, in Goodbye to All That, describes his time as an instructor and front-line officer during the First World War, observing that proficiency at parade-ground drill correlated with combat effectiveness:

> We [the officers at the Harfleur "Bull Ring"] all agreed on the value of arms-drill as a factor in morale … I used to get big bunches of Canadians to drill: four or five hundred at a time. Spokesmen stepped forward once and asked what sense there was in sloping and ordering arms, and fixing and unfixing bayonets. They said they had come across to fight, and not to guard Buckingham Palace. I told them that in every division of the four in which I had served … there were three different kinds of troops. Those that had guts but were no good at drill; those that were good at drill but had no guts; and those that had guts and were good at drill. These last, for some reason … fought by far the best when it came to a show … I told them that when they were better at fighting than the Guards, they could perhaps afford to neglect their arms-drill.[73]

In fact, drill forces the soldiers to behave in a way typical of high V, which means acting cooperatively within a controlling hierarchy.

In elite combat units military training also acts as a filter, selecting recruits who already have a reasonably high level of V, since only they will be amenable to the benefits of intense V-promoting practices and able to withstand its rigors. We have seen that V is increased by severe and intermittent stresses, such as famine or predator attack. Traditional military training is designed to inflict just such stresses. An extreme example is the Imperial Japanese army before 1945, where soldiers were not only tightly controlled but physically beaten by their superiors. In Western armies, abuse has traditionally taken the form of insults and dressing down by an NCO, but the biochemical effects would be similar. Any such experience, especially combined with a program of rigorous exercise, must elevate stress hormones such as cortisol and increase V.[74]

In summary, the principal V-promoters are illustrated in Fig. 4.1 below.

Fig. 4.1. V-promoters. V is promoted by an anxious but indulgent mother, discipline and authority in childhood, and high status plus intermittent stresses in later life.

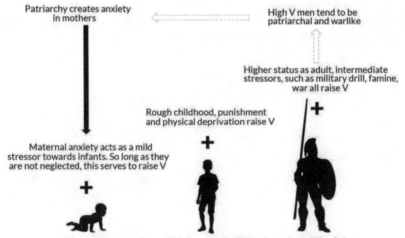

Patriarchy creates anxiety in mothers

High V men tend to be patriarchal and warlike

Higher status as adult, intermediate stressors, such as military drill, famine, war all raise V

Rough childhood, punishment and physical deprivation raise V

Maternal anxiety acts as a mild stressor towards infants. So long as they are not neglected, this serves to raise V

V promoters experienced in infancy and childhood maximise V in adults

A civilization's military prowess, and the political, cultural and economic advantages that go with it, are the most obvious and perhaps the most important benefits of high V. But there are others, all of which come under the general heading of "vigor." Widespread high morale is one. The trait that we call the "pioneer spirit" is another. High-V animals are willing to migrate in search of resources, and high-V humans have a similar boldness. At the level of the nation or tribe, this may be expressed in the drive to expand, to build an ever-bigger kingdom or empire, or to exploit new territory. In a culture that is moderate in terms of C, high V may produce migratory hordes of aggressive raiders such as Vikings or Mongols, or the waves of barbarians that repeatedly overran China and the

Middle East. In a high-C society with settled, highly productive populations and advanced economies, high V can build vast empires.

But V has another function in that it can actually support and even increase the level of C in civilized societies. To appreciate this, we must first understand that the effects of C-promoters and V-promoters depend to a great extent on the age at which they are applied, and that is the subject of the following chapter.

Testing

In a naturalistic environment as described in the previous chapter, a surge of V could be induced by subjecting an animal to stresses such as a white noise or predator odor. When nursing mothers are treated in this way, their adult offspring should be more aggressive and confident and (depending on the species) better at group co-ordination.

CHAPTER FIVE

INFANCY AND CHILDHOOD

Train up a child in the way he should go, and when he is old he will not depart from it.
—Proverbs 22: 6

Taken together, the temperamental complexes labeled C and V can be considered the fundamental building blocks of civilization. Most of the traits associated with civilization—industriousness, intellectual and technological advancement, market-based economies, and so on—are related to C. In general, the higher the level of C, the greater a civilization's achievements in those areas, and the greater its level of social, political and economic complexity.

As demonstrated in the last chapter, aggression, competitiveness and vigor are all characteristics of successful civilizations. They are grouped under the label V, and levels of C and V in a given civilization are not necessarily equivalent. A society with high C can have low V, and vice versa; or a society with high C can have high V as well. Societies with both high C and high V tend to be the most successful.

However, just as C alone does not define all differences between civilizations, because it does not account for vigor and aggression, C and V together are also not entirely sufficient. Two societies may be comparably large, wealthy, technologically advanced and militarily powerful, but have completely different forms of government—one a liberal democracy, the other an autocratic republic. Or two societies might compare closely in their general modes of government and economies, but differ greatly in their actual levels of economic productivity.

And so on, and so on. Given only C and V to work with, how do we explain differences in wealth between advanced societies as well as mechanical skills, creativity and whether a country will become a democracy or a dictatorship? There are variables in civilizations beyond the model developed thus far. We need to go further.

The answer, as with so many elements of biohistory, can be found in the experiences of early life.

As we know, the levels of C and V in people can vary significantly depending on the strength of the C and V-promoters those people are exposed to. Greater restriction of sexual behavior and tighter control of children produces a higher general level C. Likewise, harsher and more rigorous temporary stresses in childhood lead to higher V, provided they are not so severe and chronic that the "toughening effect" cannot operate.

What we find when we look closer at C is that there are significantly different effects *depending on the age in which those C-promoters are experienced.* For instance, parental control in infancy produces very different long-term effects on temperament and behavior than control in later childhood.

It will be immediately apparent that there are many potential ways in which the age factors can be combined. A person who experiences a C-promoter in infancy *and* later childhood will have a different temperament from one who experiences it only in infancy.

The ways in which behavior can be differently affected by this vary so greatly that we need to give them subcategories of their own.

We have seen that C is a product of C-promoters throughout life, from infancy through to adulthood. But *infant C* is a form of C created by C-promoters such as a tight control of children in the first few years of life, up to about age 5. Levels of C and infant C can, and often do, differ. A grown man could, for example, have high infant C but only moderate C, depending on his experiences at each age, although the highest levels of C always involve training of infants. It is also important to understand that infant C does not refer to the level of C in infants, but the strength of the C-promoters an individual was exposed to in infancy.

V also takes two different forms. We have seen that V is associated with an anxious but nurturing mother, stresses such as exposure to authority in late childhood, and occasional stresses plus high status in later life. *Child V* is the form of V associated with exposure to authority in late childhood. As indicated by the animal studies referred to in the last chapter, it is this form of V that is associated with a more tradition-minded temperament, and we will also see that people with high child V are more likely to accept political authority. This is the factor, along with the impersonal loyalties stemming from high C, that makes possible large and well-organized states.

It is important to recognize that parental control is a C-promoter as well as a V-promoter, with the relative effect depending on the method of control. Severe, punitive discipline is a C-promoter and also a strong V-promoter. A form of discipline based on encouragement and appeals to affection would be a C-promoter but only a mild V-promoter. Severe but arbitrary punishment would promote V but not C. Severe punishment in late childhood also tends to increase the overall level of stress.

The basic categorization is C, infant C, V, child V, and stress. These can combine in many complex ways. So what do these differences mean in terms of behavior and attitude, and how are they important in civilizations? We will deal with each of them in turn.

V—aggression and war

The strongest characteristic of societies with high V, whether or not they have high child V, is warlike aggression. We have seen this in the Yanomamo, a highly aggressive people with minimal political organization. Successful warfare requires not only aggression but the ability to cooperate and work as a cohesive unit, which is also characteristic of V.

Child V—larger political units, tradition

Cross-cultural studies show that the form of childrearing most strongly associated with large states is the obedience training of older children, the basis of which we have termed child V.[75] Despite their many great differences, one of the key features of civilizations such as China, Egypt, India, Japan and Western Europe is their capacity to form large states. Systems of government may be despotic, monarchical or democratic, but they govern at a level far above the local village or district.

As we saw in chapter two, the Yanomamo rarely control their children at all. Politically, they are fragmented into villages that constantly feud amongst themselves. The limited discipline of children is typical of small-scale societies. For example, Margaret Mead described the childrearing patterns of four small-scale New Guinea societies which differed dramatically in some ways. The Mundugumor were harsh and punitive and the Highland Arapesh mild and nurturing, the Manus strictly controlling of infants, but in none of these societies did parents exercise systematic discipline in later childhood.[76]

By contrast, all societies that form large states appear to discipline their older children, whether or not they control infants. But while discipline in general makes larger political units possible, the effects vary greatly depending on the relative levels of control and punishment as noted earlier. Control with little punishment produces only moderate child V. Severe punishment, with or without systematic control, leads to high child V.

Animal studies hints at another aspect of child V. A group of male rats was stressed before puberty and then conditioned to associate a sound with a shock. As adolescents, but not as adults, they showed an increase in conditioned fear, and as adults this conditioning was harder to extinguish than in rats which had not been stressed.[77] In a related study, rats stressed with a foot shock during their third week were also slower to abandon a conditioned fear in later life, something not found when experienced in the second week.[78] This suggests that stress in the juvenile period or at puberty makes conditioning easier to impose but harder to erase when animals become adults.

In human terms, we will find that people with high child V tend to be more traditional.

C—Work, business and religion

High C behavior is that epitomized by gibbons—monogamous nuclear families, limited sex, delayed breeding, intense care of young, and relentless search for food even when not hungry (the non-human equivalent of work). In human societies the highest level of C is only found in societies which control infants, but C-promoters in late childhood and adult life also increase the level of C. People with higher C tend to be hard working, sexually restrained, modest in their habits and dress, and somewhat more introverted and lower in testosterone than those with lower C.

Europeans and Japanese of the nineteenth and early twentieth centuries are all examples of societies with high C. However, within most high-C societies there are subgroups that have even higher C than the average. A particularly telling example of this can be found in Mormons.

Devout young Mormon men subject themselves to an extraordinary degree of austere discipline while engaged in a two-year Mission. They are prohibited sex and even dating, secular music, and most forms of entertainment. As well as evangelizing for their religion, they live a quasi-monastic life dedicated to prayer and theological study. They are also socially disciplined, deferring to their elders.

All devout Mormons, including those who are married, are required to practice strict sexual restraint. The level of C promoted by such a life is much higher than that found in the wider societies in which Mormons live. Therefore, it is not surprising that Mormons are successful out of proportion to their numbers. Mormons comprise barely 2% of the American population but lead a number of major corporations including JetBlue, American Express, Marriott, Novell, Deloitte and Eastman Kodak. There are more than a dozen Mormons in Congress including Senate Majority Leader Harry Reid. Mitt Romney and John Huntsman were prime contenders for the White House.[79]

The Jains in India likewise have higher C than average for the nation, which is reflected in their disproportionate economic success. Despite being only 0.5% of the population, they pay a quarter of India's income tax.[80]

A distinct feature of high C is the willingness to sacrifice present consumption for future benefit, which is a key requirement for occupational success. Parental control in late childhood promotes C, and so too does restriction of sexual activity. This last applies especially in adolescence, as shown by the Kinsey report cited in chapter three. Kinsey demonstrated that people with lower than average sexual activity in the crucial years after puberty attained significantly more education and career success.

A high level of C also plays a significant role in creative and intellectual achievement, partly because it helps to increase focus on a subject for long periods at a time. It may also make people more flexible in their thinking. For creativity, the ideal combination would be intense control in early childhood, maximizing C, but also the "flexible" infant C. Discipline in later childhood could also be a positive, provided that it depends principally on control rather than punishment. In this case, any moderate increase in the "traditional" child V is more than balanced by the work ethic and concentration made possible by higher C. Punitive discipline without systematic control, on the other hand, promotes traditional thinking with less of a C-promoting effect.

C-promoters such as sexual restraint in adult life should also promote creativity by increasing C without any significant effect on child V. And it is to this subject that we must turn now.

Malcolm Gladwell's book *Outliers* proposes that outstanding success in any field requires at least 10,000 hours of dedicated practice, a compelling case which draws on examples as diverse as Bill Gates, the Beatles, and ice hockey players.[81] This is exactly the kind of concentration that is favored by very high C. His case studies also give clues as to how such very high C might be achieved, in that it is striking how often people of genius appear to have had very limited sex lives. In some cases this may have been because their preferences were outlawed by society. For example, Leonardo da Vinci never married and was at one time charged with sodomy.[82] Michelangelo's paintings and love poetry also suggest he was homosexual by orientation.[83] Isaac Newton never married and was commonly supposed to have died a virgin.[84] Stephen Hawking was known as a brilliant but lazy undergraduate, and only began to focus and achieve his outstanding insights after he developed motor neuron disease, a condition known to limit sexual activity (though Hawking was still able to marry and sire three children).[85]

Many people throughout history have been open to new ideas, which in most cases simply makes it easier for them to abandon traditional beliefs and practices. One traditional belief very likely to be abandoned by such people, because it is so at odds with temperament, is that concerned with limits on sexual behavior. Thus, a person with a flexible mind set is far *less* likely than most to attain the highest level C as an adult. It is only when flexibility of thought is combined with severe limits on sexual behavior (including the solitary kind), and linked with high intelligence and opportunity, that true genius is most likely to arise. This helps to explain why it is so rare.

Infant C—industrialization, creativity and wealth

Control of older children is a near universal of civilized societies, but control of younger children and especially infants is not. In fact, there is only one major culture area where people came to control infants as tightly as they did older children—Europe.

Very young children in nineteenth-century Europe were under constant and extremely close control. The archetypal embodiment is the English nanny, who was responsible for raising virtually all infants in upper-class families. A prime example of how they enforced their renowned discipline, and the kind of values they represented, is their obsessive drive against sexual behavior, and their rigorous and very early administration of toilet training:

> Pot training began very early, in the first month or so. Thereafter it was continued, in a sense, right up to the moment the child left Nanny's care … Training took several forms. At first—innumerable placings on the pot, dozens a day. Once the baby or child had learnt, the most common method of persuading a child to go was to leave him or her there … till something "happened." Eleanor Acland describes in her autobiography how "It was nursery law that we might not quit the water closet till we were fetched." Once one of them was forgotten from tea until bedtime and sat there for five hours.[86]

Based on parental advice books, such rigorous training also seems to have been commonplace for the literate middle classes. And this was not just the case in England. For example, the most authoritative Dutch family manual of the Victorian era was *The Development of the Child* by Dr Gerard Allebe. Dr Allebe's main concern was that mothers not be sentimentally over-protective. This was a serious danger because it would deprive the child of the necessary exercise for courage and result in weakness. Mothers must suppress their inclination to indulge and spoil their children, as doing so would result in the children becoming fearful.[87]

Swedes seem to have focused even more on the early years, as judged by *The Century of the Child*, published in Sweden in 1909 and becoming an immediate bestseller.[88] The idea seems to have been that if discipline was firm and consistent in infancy it would hardly be needed later:

> Only during the first few years of life is a kind of drill necessary, as a pre-condition to a higher training. The child is then in such a high degree controlled by sensation, that a slight physical plain or pleasure is often the only language he fully understands. Consequently for some children discipline is an indispensable means of enforcing the practice of certain habits. For other children, the stricter methods are entirely unnecessary even at this early age, and as soon as the child can remember a blow, he is too old to receive one … The child must certainly learn obedience, and besides, this obedience must be absolute. If such obedience has become habitual from the tenderest age, a look, a word, an intonation, is enough to keep the child straight … With a very small child, one should not argue, but act consistently and immediately.

In these countries and exactly at this time there was an explosion of economic activity involving machinery, for which people at the time seemed to have an extraordinary talent. Nineteenth-century Europeans also developed highly efficient governments based on impersonal institutions such as parliaments and the rule of law. In the next chapter we will trace the rise of this childrearing pattern over several centuries.

Could there be a connection between this extraordinary flowering of technology, science and democracy and the extraordinary and unprecedented level of control of infants? One clue can be found in the fact that cultures which did *not* industrialize in the eighteenth and nineteenth centuries, and which were as strict as Europeans with older children, were completely indulgent of infants. Europeans found this very odd, as noted by one visitor to China at the close of the nineteenth century:

> He is welcomed to the household with a wild delight, to which it is wholly impossible to do justice. He begins life on the theory that whatever he wants, that he must have; this theory is also the one acted upon by those who have him in charge, to an extent which seems to us ... truly amazing. A Chinese mother is the slave of her children. If they cry, they must be coddled, most probably carried about, and at whatever expense, if it is possible to prevent such a terrible thing. They must not be allowed to cry continuously.[89]

Serious discipline began much later, around the age of 5 or 6:

> Taiwanese parents assume that children cannot really "understand" until they are around six years old. They claim that until then they do not try to teach them anything and expect little in the way of obedience. At most they hope that the preschool child will not injure herself or cause her parents too much trouble. [90]

Egyptian practices, well into the twentieth century, were not substantially different. Infants were indulged, fed on demand, and not disciplined or controlled in any way.[91] Unlike Britain and Western Europe, neither China nor Egypt industrialized early, and remained poverty-stricken and economically underdeveloped well into the late-twentieth century. As we shall see, this is probably no coincidence.

If the early control of infants drives early industrialization, then we should be able to observe this pattern in other parts of the world too. One particularly interesting example is Japan, which is the only non-Western nation to industrialize relatively early (though not as early as Europe).

The process began in the late-nineteenth century, from around 1870 onward. The Japanese elite, recognizing the power and wealth of Western nations, made a conscious decision to industrialize. Western science was adopted, along with European methods of education, and there was massive investment in railroads and industry.

Although it succeeded in building itself into an advanced industrial nation, Japan has to be considered a successful imitator, not an originator, of industrial development. Therefore, if our hypothesis holds, we should see a higher level of infant control than in the Middle East and China, though not to the same level practiced by Europeans in the nineteenth century. And sure enough, this is exactly what we find. As observed in the Japanese town of Niiike in the 1950s:

> Until well into the toddling age, a child "does not understand" attempts to train him, so Niiike people do not hold it against a youngster if he fails to follow instructions nor do they blame him for disobeying. Actually, of course, many principles and habits are instilled in the baby before the age of "understanding," and the real consequence of the Niiike point of view ... is a minimum of early punishment or negative discipline.[92]

Though far less rigorous and severe than Victorian nannies, Japanese parents in this community began training their children well before the age of two. The same pattern can be seen in a remote rural village in the 1930s. Toilet training was typically complete in the first year of life and polite manners taught from infancy by constant repetition and instruction.[93] This early start is especially striking given that these parents were far less strict than in Niiike, in that younger children would commonly disobey their parents.

The key difference, as mentioned earlier, is that the Japanese training of infants seems to have been far less rigorous than it was in nineteenth-century Europe or America, which would indicate a level of infant C higher than in most countries but lower than the West. Thus it is that Japan, though it developed an advanced market economy and high levels of literacy before opening to the West in the mid-nineteenth century, never industrialized of its own accord.[94] Only under external pressure did it make the changes that turned it into an industrial power within fifty years.

Another feature of Japanese history in common with Western Europe is that the modernization of the economy went hand-in-hand with the early development of democracy. Modern Japan and most European nations are able to operate as functional democracies with relatively low levels of corruption. It is possible for nations without the early control of infants to become democratic (India being one obvious example), but on the whole it is far easier with high infant C.

It seems that infant C, promoted by the early control of infants, correlates and is likely a major driver for industrialization, science, intellectual creativity, the capacity for efficient democratic government, and an advanced market economy.

The idea that industrialization can be attributed to the control of infants is a radical one but, like most of biohistory, not difficult to test. One way would be to interview the parents of young people with advanced mechanical and engineering skills, asking them to recall how they treated their very young children. Another way would be to find a society that controlled infants but *not* older children. It should have features distinguishing Western and Japanese societies from other advanced civilizations, such as openness to change and an unusual aptitude for machines, but without the large political units associated with child V. It just so happens that such a society did exist, if only for a brief period, and we will consider it later.

Nation States versus Empires—child V with and without infant C

While all societies which discipline their children in late childhood tend to form large states, the level of early control (infant C) seems to affect the kind of state. During the nineteenth century in Western Europe, civilizations tended to coalesce more and more into individual nation states characterized by single cultures. Each nation had a single language and a single cultural and ethnic identity, or would at least be dominated by the most common one. The reason is that people with high infant C seem to be extraordinarily resistant to rulers of different cultural and ethnic groups. An early example is the bitter struggle of the Dutch for independence from Spain in the late sixteenth and early seventeenth centuries. At that time Spain was the pre-eminent super power of the day with the vast wealth of the Americas at its disposal, but the Dutch were in the end successful. An even more striking example is Poland, divided among neighboring powers in the late eighteenth century but emerging nearly two centuries later with its language and culture intact.

Outside of Europe, however, in places such as the Middle East, China and non-British India, vast and cosmopolitan empires were formed which expanded to the entire cultural area and beyond. Boundaries were set by physical obstacles such as mountains, deserts or the sea. Even though India, China and the Middle East were each comparable in size and population to Europe, none of them developed stable nations *within* the culture area, with post-colonial states set up under Western influence being the one exception. Britain and other Western nations were quite capable of being imperial powers, in the sense of controlling other peoples, but stubbornly resisted becoming part of anyone else's empire.

We can now add another aspect to our understanding of infant C. Child V, the result of discipline in late childhood, causes people to accept wider political authority and thus form large states. But when it is combined with infant C, resulting from control in early childhood, people are only ready to accept the authority of rulers who are similar in language and culture, and fiercely resist more alien rulers. Thus, along with the capacity for early industrialization, creativity, democracy and advanced economic systems, civilizations which practice control of infants also tend to form nation states.

We have seen that C-promoters in adult life, such as sex restraint, can have effects different from control in later childhood because they do not involve an increase in child V. It is also possible that the effects of C-promoters in adult life are distinct in other ways, an effect that could be termed 'adult C'. Behaviors especially linked to adult C could be a greater interest in children, and religious fervor. People brought up strictly tend to be hard working and disciplined, but are often secular in their outlook and not especially fecund. This is something that will be the subject of further research.

Stress

Child V is increased by control or punishment in late childhood, but as noted earlier a great deal depends on the relative level of each. The Japanese community cited earlier was patriarchal and thus high V, but control of children was associated with very little punishment.

> Loud-voiced commands, repetitive and detailed instructions, scolding or tongue lashing, and physical beating are relatively rare and disapproved of in Niiike homes ... A mother angered at her child may pinch him painfully. A father pushed beyond endurance will cuff the child on the

head, but only rarely. Physical contact is used more often as a positive method of instruction than as punishment.[95]

Such an upbringing produces only moderate child V, so status differences are milder and there is less fear of authority. Contrast this pattern to that of a community in Egypt, where punishment of children was both random and brutal, producing not only higher child V but a higher level of stress:

> Connected with producing fear in the children is the violent and bad-tempered manner in which adults administer punishment to them. Punishment ... may be in the form of fulminations or curses, or it may be corporal ... Corporal punishment is not uncommon either by beating, striking, whipping or slapping ... In administering punishment there is no consistency or regularity; for the same offense the child might be beaten harshly, or his offense allowed to pass unnoticed.[96]

Or China:

> A beating administered by a Taiwanese parent is often severe, leaving the child bruised and in some cases bleeding. Parents prefer to use a bamboo rod to discipline children, but they will use their hand or fist if there is no bamboo available, and if they are really angry, they will pick up whatever is at hand. Crueller forms of physical punishment are also used by a few parents, such as making the offending child kneel on the ridged surface of an abacus or tying the child in a dark corner ... Mothers commonly punished children in a violent fit of anger, so uncontrollable at times that family members and even outsiders might intervene to prevent the child from becoming seriously injured.[97]

Cross-cultural evidence suggests that punishment in late childhood is associated with authoritarian government, involving huge differences in power between rulers and ruled.[98] The reason for this can be seen in the Egyptian, villagers described in chapter one, who respected only powerful and brutal authority.[99]

It was this acceptance of harsh authority, combined with the lack of national loyalties associated with low infant C, that has made possible the powerful and extensive Empires of the Middle East until recently. It also explains why the American invasion of Iraq, and the "Arab spring" which removed brutal tyrants across the Middle East, failed to bring about peace and democracy. On the whole the result has either been continued anarchy as in Iraq, Syria and Libya, or a return to authoritarian rule as in Egypt. Fear is the only effective rule for stressed societies. Or, in other words, governments reflect society rather than vice versa. This is a core tenet of biohistory.

Societies that punish children in late but not early childhood are more traditional because they lack infant C and have higher levels of child V. When control is less systematic and there is none in early childhood, they also have less of the mental flexibility associated with C.[100] In this Egyptian, village there was deep resistance to change. Religious belief and behavior were pervasive, governing all aspects of life, and social patterns and customs had changed little despite decades of government with modernizing ideas.[101]

The Manus

We have suggested that large political units are promoted by child V, while machine skills and innovation—key characteristics of nineteenth century Europe and Japan—are products of infant C. A perfect test of this theory would be a society which controlled infants but *not* older children, which would produce infant C without child V. Such a people would be innovative, good with machines and strongly oriented to the market, but would not tend to form large political units.

Fortunately, such a society did exist, if only briefly. This is the Manus people of the Admiralty Islands, as described by Margaret Mead in the 1930s. At this time they were living in stilt houses over a lagoon, an exceptionally dangerous environment for infants. This made some rigorous training necessary to prevent them falling in the water:

> When [the child] is about a year old, he has learned to grasp his mother firmly about the throat, so that he can ride in safety ... The decisive, angry gesture with which he is reseated on his mother's neck whenever his grip tended to slacken has taught him to be alert and sure-handed ... For the first few months after he has begun to accompany his mother about the village the baby rides quietly on her neck or sits in the bow of the canoe while his mother punts in the stern some ten feet away. The child sits quietly, schooled by the hazards to which he has been earlier exposed.[102]

But training went far beyond the requirements for physical safety. It also applied to property:

> Before they can walk they are rebuked and chastised for touching anything which does not belong to them. It was sometimes very tiresome to listen to the monotonous reiteration of some mother to her baby as it toddled about

among our new and strange possessions: "That isn't yours. Put it down.
That belongs to Piyap [Mead]. That belongs to Piyap. That belongs to
Piyap. Put it down." But we reaped the reward of this eternal vigilance: all
our possessions … were safe from the two and three-year-olds who would
have been untamed vandals in a forest of loot in most societies.[103]

Young children were also taught rigorous privacy in excretion and a strong
sense of modesty. But this rigorous training of infants was accompanied
by an almost total lack of discipline or obedience after infancy.

> The children are taught neither obedience nor deference to their parents'
> wishes. A two-year-old child is permitted to flout its mother's request that
> it come home with her. At night the children are supposed to be home at
> dark, but this does not mean that they go home when called. Unless hunger
> drives them there the parents have to go about collecting them, often by
> force. A prohibition against going to the other end of the village to play
> lasts just as long as the vigilance of the prohibitor ….[104]

Manus society showed exactly the characteristics we would expect from a
society with high infant C and low child V. Consistent with low child V is
that there was no political unit beyond the village, even though all
communities had originated in a single village in recent times. Local
power was based only on wealth and the abrasive, forceful personality
brought about by wealth and power.

Consistent with high infant C is that the Manus were highly successful
traders, obsessively hard working and ambitious and collecting a great
deal of wealth. Underlying the fervent capitalism of the Manus was an
intensely moralistic religion, based on ancestral spirits, and focusing very
heavily on property. And just as the infant C societies of Europe have been
less prone to corruption than many third world societies, so the Manus had
a rigid and rigorous sense of commercial honesty.

They also, and very significantly, showed an unusual aptitude for
machinery, an enthralled fascination that had them examining and
tinkering with every engine and gadget they could find, to the vast
amusement of American troops stationed on Manus island during the
Pacific War. Commenting on the American machines later, after the war:

> We can understand Australian engines … easily because they are all open
> and you can see them work. The American engines are closed up in boxes,
> and all you can see is the button that starts it and the button that stops it.
> But, they added confidently, if we could just see inside, we would
> understand how it works.[105]

A final and significant feature of Manus society was openness to change, once again reminiscent of the flexibility of the modern West. This was shown not only by their success as traders, which requires a great deal of flexibility and creative thinking, but their extraordinary willingness to adopt—almost overnight—an ideology that reversed almost every tenet of their culture. Margaret Mead describes their obsession with change and new ideas:

> The great avidity with which they seized on the new inventions which came with European contact, was partly rooted in their driving discontent with things as they were.[106]

Within a few decades they completely abandoned their religion, their economic system, and their ideas of social relationships in favor of a radical cult called the "New Way."

The combination of very high C with low child V makes this case study extraordinarily interesting. It indicates that high C does not necessarily result in the training of children at all ages. Tight control of children is not possible without high C, but C directs behavior in ways specified by the culture. Usually, cultural values support the training of children, but the Manus case indicates that it need not.

Another useful aspect of this case study is that there was a generation of Manus who passed late childhood and adolescence during the Japanese occupation, and so had a much more stressful time. This cohort, with higher child V, was more accepting of authority. There are descriptions from this time of people listening for hours to the cult leader, Paliau, without "batting an eyelid."[107] They were also more rigid and resistant to change:

> They had had, in fact, the same early childhood as the older men, but had lacked the kind of late childhood and adolescence in which the older men's habits of companionship and friendliness, and their capacity to feel free, had been born. They were stiff and difficult fathers, with less tenderness and indulgence to their children than the older or much younger men … In the brittle, dogmatic orthodoxy of these young men, there is limited refusal to accept anything new.[108]

This confirms the idea that the effects of child V are opposite to those of infant C, making people less open to change rather than more so. It is a notion which helps to explain why societies with high infant C and low child V are so rare, even though the Manus people were hugely successful in their time. Such a society is extremely ready to abandon its traditions, which is not a good recipe for cultural longevity.

Summarizing, then—the variants of C, V and stress can be defined as having the characteristics listed in Table 5.1 below.

Table 5.1. Variants of C and V. These are the characteristics of adults who experienced C-promoters, V-promoters and stress at different periods of life.

C	All ages	Industrious, disciplined
Infant C	Control to age 5 but especially 0–2	Flexible, machine skills, prefer similar rulers
V	Anxious mothers, occasional stresses, high status as adults	Warlike
Child V	Authority and stress at age 6–12	Traditional, accepts authority
Stress	Punishment at all ages	Fear and stress

Testing

As discussed in chapter two, rats whose mothers were food restricted show epigenetic settings as adults that could measure the level of infant C. Because these are found in the brain they are not a practical method for assessing infant C in humans, which is why we are currently looking for a way of assessing this through blood tests. Once developed, people with these settings should be better mechanics and engineers and have other behavioral indications of high infant C.

CHAPTER SIX

THE RISE OF THE WEST

Nothing that is morally wrong can be politically right.
—William Gladstone

In the previous chapter we saw how experiences in early life affect character, and that methods of childrearing can explain the political and economic makeup of a society.

C is increased by parental control in childhood, by sexual restraint, self-discipline, and moderation in personal habits such as eating. People with high C tend to be hard working, religious and to favor occupations requiring intense focus and long-term orientation, such as business and the professions.

A high level of infant C, which is found in adults subjected to strict control in early childhood, orients people to the market economy and is associated with mechanical aptitude. People with high infant C prefer rulers who share their language and culture, so societies with high infant C tend to form nation states (or, as will be noted later, city states) rather than cosmopolitan empires.

V is promoted by certain forms of stress including anxious but nurturing mothers in infancy and intermittent stresses in later life. It is undermined by neglect or abuse in infancy and chronic stress thereafter. People with high V tend to be aggressive, confident and energetic, and to co-operate well within small groups.

Child V is promoted by experience of authority and control in late childhood. It makes people more accepting of authority, and also more conservative. When applied as punishment, increasing the level of stress, adults more readily accept autocratic and hierarchical regimes. Assuming that they have not been severely stressed in infancy, people punished in late childhood tend to be especially traditional.

If the model developed so far is broadly correct then it should be possible to analyze the development of a civilization using these variables alone. More specifically, the rise of the West can be explained by an unprecedented rise in the level of C, and especially infant C. Temperament determines the attitudes and behaviors upon which society is built, such as attitudes to law and authority, levels of mechanical and economic aptitude, preferences for different forms of social and political organization, and tolerance of moral and cultural change. Changes in C and V can help to explain all of these. The rise of C explains not only major political and economic changes that have taken place in the West over the past thousand years, but also changes in family and childrearing patterns that mirror what happens to animal societies when food is restricted.

As an explanation for the rise of the West, biohistory is one of many.[109] As a theory of history whose conclusions can be rigorously tested in the laboratory, it stands alone.

Changes in sexual attitudes and behavior

Sexual behavior is one of the most obvious indicators of a society's level of C. Just as gibbons in food-restricted environments limit their breeding, human attitudes toward sexuality are determined by the level of C.

During the earlier Middle Ages, sexual morality was far from rigorous. Women were regarded by churchmen and medical practitioners as more innately lustful than men. Dictates of the church aside, expectations of sexual behavior among most of the population were decidedly relaxed. Pre-marital sex was fairly common and did not necessarily harm marriage prospects.

But from the twelfth century onward, standards began to tighten. The influence of churchmen—their own C raised to a pitch by monastic and other disciplines—brought about a general rise in the C of the population. Sexual prohibitions became stricter and more ingrained. During the twelfth century the church had freed itself from secular control and became a form of self-governing theocracy. Its independent power—and the power it exercised through monarchs—gave it license to impose its own standards on all levels of society.

It was during this period that clergy were forbidden to marry, and divorce outlawed.[110] By the fourteenth century, loss of virginity in an unmarried woman was seen as an offense against God rather than merely an affront to family honor. Though still, at this stage, punishments for pre-marital sex remained rare.[111]

Attitudes became much tougher during the Protestant Reformation, and by the nineteenth century had reached an unprecedented level of prohibition and control. The British middle class worked hard to discourage masturbation, then regarded as a serious illness. From the late eighteenth century onward, children's literature warned that it could cause blindness or madness. This is not true, of course, but curbing early sexual activity does aid educational and occupational success, as indicated by the Kinsey Report discussed in chapter three. C-promoting beliefs need not be true to be effective.

Some professionals went so far as to argue that no one should become sexually active before the age of 25. All manner of sexual activity outside marriage was severely proscribed, and even sex within marriage was commonly seen as an undesirable necessity to be kept under tight control. Men should only have sex with their wives to reproduce. To do otherwise was to treat her like a prostitute. Most significantly, however, was that respectable women were believed not to experience sexual desire. As Dr William Acton wrote in 1857:

> Having taken pains to obtain and compare abundant evidence on this subject I should say that the majority of women (happily for them) are not very much troubled with sexual feelings of any kind. What men are habitually, women are only exceptionally ... there can be no doubt that sexual feeling in the female is in the majority of cases in abeyance ... and even if roused (which in many instances it can never be) is very moderate compared with that of the male ... As a general rule, a modest woman seldom desires any sexual gratification for herself. She submits to her husband, but only to please him; and but for the desire of maternity, would far rather be relieved from his attentions.[112]

It had taken a thousand years, but the attitudes of the medieval church had finally conquered civil society. This increase in self-discipline and sexual restraint both reflected and drove the immense rise in C, and especially infant C, that peaked around 1850.

This rise in C, like the increase in strict sexual morality, was not the same in all social classes. The middle classes were the most restrained and showed the greatest rise in C. Elements of the working classes, on the other hand, remained freer and more open. There were many more illegitimate births among them and it was not unknown for couples to engage in intercourse before their wedding. But members of the respectable working class were influenced by more rigorous standards of sexual morality, and were considerably more restrained than their medieval ancestors.

This not only reflects a rise in C of the general population, but a rise to cultural and political dominance of the high-C middle classes, as the political institutions and market economy that their temperament favored grew in size and power.

The changing age of puberty

Another variable associated with C is the age at puberty. Higher C people and animals reach puberty later, reflecting a biological function of C which is to delay breeding in food-restricted environments. The declining age of puberty over the past century and a half is commonly attributed to improved nutrition, but this does not explain why it rose so much over the previous five hundred years.

In medieval Europe, doctors placed the typical age of puberty at around 12 to 13, roughly what it is today. In the fourteenth century, boys were considered sufficiently adult to be subject to poll tax at the age of 14, an age indicated by the presence of pubic hair. But over the centuries the average age of puberty rose, and by the nineteenth century it was as high as 17 in some European countries.

Many explanations have been given for this increase. The so-called Little Ice Age from 1300 to 1850, when temperatures fell and farming became less productive, is sometimes used as evidence for poorer nutrition. But the fourteenth century suffered widespread famines and yet puberty was as early as it is today. Also, skeletal evidence shows that people in England were about as well-fed in 1800 as they had been in 1200.[113] Nutrition alone does not explain delayed puberty, but the C system does. The increasing activation of this underlying physiological system, which mimics the effects of food shortage, had the effect of delaying puberty.

Later age of marriage

Late marriage is the cultural equivalent of delayed puberty. In high-C societies, where puberty is later, marriage tends to be delayed until well after the age of puberty. Once again this has a parallel in the behavior of gibbons, which breed only after securing a territory, normally well after puberty.

In medieval Europe, marriage in early adolescence was common even among commoners. Among the nobility children were often married in childhood for political reasons, though the marriage was not normally consummated until puberty.[114] Margaret Beaufort, for instance, gave birth to her son (the future Henry VII) in 1457 when she was only 13. But by the eighteenth century the age at marriage had risen significantly, and a widening gap had opened up between the onset of puberty and marriage. In the nineteenth century, puberty occurred at about 17 and marriage at about 24 (see Fig. 6.1 below).

Fig. 6.1 Age of first marriage for women in royal and ducal families.[115] This shows a steady rise to the nineteenth century, consistent with a steady rise in C.

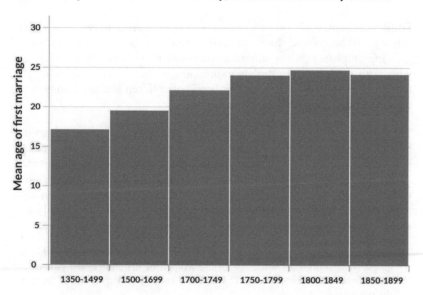

There were class differences. Parish records indicate that among the working class, the age of marriage rose considerably but peaked in the eighteenth rather than the nineteenth century.[116] The reasons for this will be discussed in a later chapter.

The Nuclear Family

Nuclear families based on monogamous pair bonds are one of the most noticeable features of gibbons and other high-C animal populations. Humans have a species bias towards pair bonds so that even in low-C populations (such as hunter-gathers) most marriages are monogamous, but monogamy is *more* common and polygyny *less* common in high-C societies. High-C human societies also show an increased separation between the monogamous couple and the rest of society.

During the Middle Ages, although monogamous marriage had long since been the norm, family groupings were much larger. Extended families tended to live together in large, multi-generation groups. Social activities, particularly festivals, were also far more communal than they later became.

From the fourteenth century onward people increasingly valued their privacy. Between the fifteenth and nineteenth centuries there was a growing emphasis on the intimate emotional bond between husband and wife. This achieved its ultimate expression in the Victorian middle-class family. Husband and wife and their brood of children formed a household, rarely intruded upon even by close kin.

The concept of family had shifted from a primarily economic unit to one bonded by sentimental, emotional ties.[117] Relationships with more distant kin—grown-up siblings, cousins, grandparents—declined in importance. At the same time, the divide between servants and their employers grew wider.[118] Although no human culture has ever practiced gibbon-style territoriality, by the nineteenth century the ideal was that the family live alone:

> Every Englishman has, in the matter of marriage, a romantic spot in his heart. He imagines a "home," with the women of his choice, the pair of them alone with their children. That is our own little universe, closed from the outside world.[119]

The inclusive social activities of the Middle Ages, in which all classes engaged in communal festivals, gave way to a system that was structured by barriers. There was a social barrier between the family home and the wider society, with elaborate systems of calling cards and appropriate times for visits, all of which strictly regulated the families' social interactions with each other. It was now that Christmas evolved from a communal event into the intimate family occasion of modern times.

Even religious concepts began to reflect this temperamental and psychological shift towards the nuclear family. The concept of the afterlife as a heavenly host gave way to a cosy paradise where the departed would be reunited with their loved ones and live in eternal domestic bliss. By the mid-nineteenth century, this was the dominant popular view of heaven.[120]

All these factors, from the emotional preference for the nuclear family to the desire for seclusion from others, provide clear evidence of exceptionally high C in the nineteenth century.

Inhibition and reserve

Another well-known aspect of the Victorian temperament is its extraordinary level of inhibition and reserve. It was in this era that the British male acquired the reputation for reserve that persists today. This was not always so. Englishmen in the sixteenth and seventeenth centuries were said to be quick tempered, liable to impulsive behavior and openly passionate. When visiting London in the sixteenth century, the Dutch scholar Erasmus noted the emotional flamboyance of the English: "In a word, wherever you turn, the world is full of kisses."

A study of diaries from the sixteenth to the nineteenth century indicates a trend from a melancholy Englishman—quick to tears over anything from the fate of the King to a change in residence—towards an overarching sense of optimism and cheerfulness by the mid-nineteenth century. When once it was normal to refer to one's past with "great sorrow and dolefulness," and frequently to burst into tears unembarrassedly, the disposition of men had changed by the nineteenth century where cheeriness became nearly a religious obligation, and a man crying became a serious *faux pas*.[121]

By the late eighteenth century the economically and politically powerful middle classes began to constitute the British temperament we are more familiar with. Control and reserve became paramount, developing in the nineteenth century into the legendary concept of the British "stiff upper lip."

Victorians withdrew into their nuclear families. When they did indulge in communal social events they tended to be small and quiet soirées and dinner parties, rather than the large, crowded and noisy balls of the eighteenth century. By the Victorian era, the Englishman was typified as "Strong and silent, earnest, matter of fact, sparing in his speech, scornful

of theory, unemotional, manly and athletic."[122] After the horrors of the
Crimean war, stoicism became a matter of national pride. David Newsome
commented on the change in attitude as observed in English public
schools:

> [From the 1860s onward] excessive displays of emotion came to be
> regarded as bad form: patriotism and doing one's duty to the Empire
> became the main sentiments which the new system sought to inculcate.[123]

It is notable that none of these changes has a direct relation to
extraordinary economic and social progress occurring at the time, yet if we
understand C the link becomes more obvious. As with many indicators,
there was a considerable divide along class lines, with the working classes
being far more open and gregarious than the generally sober and inhibited
middle class.

Once again, this aspect of high C has parallels in animal behavior.
Research shows that animals in food-restricted environments tend to be
timid and socially inhibited.

Control of infants and young children

The control of children is a core feature of high-C cultures. It is a powerful
C-promoter and, as detailed in the previous chapter, can have different
effects depending on the age at which it applies. Control in early
childhood and especially infancy (infant C) is associated with creativity
and machine skills, while control in late childhood produces a
temperament more morally conservative and accepting of authority.

From the Middle Ages through to the Victorian period, control of children
became ever more rigorous. Equally important, from our point of view,
was a change to the ages at which it was applied. Medieval childrearing
could be severe at times, but control was not exercised before the age of 3.
Between the sixteenth and seventeenth centuries, training of children not
only became stricter but changed in character, focusing on improving the
child's moral integrity rather than simply on obedience and respect. But it
was only from the late-seventeenth century that very early training started
in earnest. Susannah Wesley gave her views on childrearing from the
early-eighteenth century:

> When turned a year old (and some before), they were taught to fear the
> rod, and to cry softly; by which means they escaped abundance of
> correction they might otherwise have had; and that most odious noise of

the crying of children was rarely heard in the house; but the family usually lived in as much quietness as if there had not been a child among them.[124]

Victorian nannies represent the height of such practices, reflecting not only middle-class attitudes but the practices of the respectable working classes from which they were drawn. Nannies in Victorian homes enforced elaborate systems of rules and regulations, often beginning in the very first month of life. Neatness and politeness in middle-class homes were taught to a degree not seen before or since. These behaviors indicate a deep-seated need to control and direct which is the essence of very high C.

Impersonal loyalties and the nation state

All aspects of C considered so far—age of puberty, family, and control of offspring—have parallels in food-restricted animal societies. But strictly human spheres such as law and political loyalty can only be linked to animal behavior by inference. The connecting notion, as discussed in chapter three, is *impersonality*.

Gibbons have intense personal bonds with their mates and immature young, but less to other members of the species. It might be said that their *external* relations are governed not so much by personal ties, as in a baboon troop, but by attachment to a section of forest. This emotional attachment to territory is by nature impersonal, and is crucial because an individual's political loyalties and participation in the market are defined by the attitude to people *outside* the nuclear family.

Likewise, as C and especially infant C increase in human cultures, loyalties become increasingly impersonal. The clan leader or local lord is no longer the focus of loyalty, which can now shift to more distant leaders. In a sense, the civilized nation is a gibbon territory writ large.

Higher C does not necessarily mean larger states, because larger political units are a product of child V. The largest European Empire in post-Roman times was that of Charlemagne in the late eighth century, which controlled what is now Germany, France and northern Italy. This later broke up, and Germany remained fragmented into a large number of micro states until the nineteenth century.

But as the level of C increases and attitudes become ever more impersonal, central control within states becomes stronger. Ultimately, if infant C is high enough, loyalties may shift to a republic or constitution where the form of law becomes more important than loyalty to any specific individual. Table 6.1 below summarizes the changes in political loyalties as C increases.

Table 6.1. Political systems connected to different levels of C. As C increases, people transfer loyalties to more distant figures and finally to impersonal concepts such as political parties and constitutions.

C	Loyalty to:	Example
lowest	Face to face	Hunter-gather band
	Local leader	Eleventh-century France (feudal anarchy)
	More distant leader	Twelfth-century France (powerful Dukes)
	King	Sixteenth-century France or England
	Representative Body	Eighteenth-century England
highest	Political Party, Law, Constitution	Nineteenth- and twentieth-century England or America

Although it is possible for cultures with low infant C to be democratic, as in many Third World countries today, this is usually due to outside influence. Democracy in such societies also tends to be unstable, undermined by personal connections that disregard principles of law (nepotism and corruption), soldiers more loyal to their commanders than to the constitution (military coups), and personal loyalty to the head of state (authoritarianism).

It is normally only within societies with high infant C that democracy arises naturally because it matches the temperament of the people. Preference for impersonality is not the only infant C characteristic that favors democracy. Infant C produces a higher level of integrity within any given rule-based system, which translates into a more efficient, effective democracy. For instance, India inherited democratic ideals from Britain but is far more prone to corruption than democracies with higher infant C, such as Japan.

The preference for impersonality is not confined to forms of government. It also affects attitudes to social order and the law, resulting in nineteenth-century European nations with relatively small and efficient bureaucracies.

Rising C also explains changes in national boundaries, which were subject to radical change in the Middle Ages. For example, English rulers at times controlled swathes of what is now France, and territories throughout Europe changed hands frequently as a result of wars, alliances and royal successions. These shifts reflected the tolerance of people with lower infant C for leaders of a different culture or language.

But as C rose, national boundaries solidified. For one European nation to conquer and incorporate another country became increasingly difficult due to the development of national loyalties. This trend was most advanced in northern and western Europe. Eastern Europeans were for a long time more tolerant of "foreign" rule, as in the case of the Austro-Hungarian Empire. But even here, peoples such as the Poles maintained a stubborn cultural loyalty through centuries of subjection.

The eighty-year Dutch war of independence against Spain during the sixteenth and seventeenth centuries shows how ferociously people with high infant C will resist foreign rule, even against extraordinary odds. Holland was a tiny maritime republic and Spain a super-power flush with American gold, but the Dutch could not be defeated. By the nineteenth century, nationalism was at an all-time high, though by 1914 it had taken on a particularly militant form for reasons to be explored in a later chapter.

Rise of the market economy

All variants of C give people a preference and aptitude for market economies, but infant C is the most market-oriented of them all. The linking factor is, again, impersonality. In pre-civilized cultures, economies are based on personal contact between individuals who know each other. For example, indigenous aboriginal societies share their resources and have loose, collective notions of ownership. The main factor determining whether a person has access to a commodity is their personal relationship with the person who controls it. For example, among the Mbuti the meat from a hunt is divided according to principles of relationship and mutual help.[125]

In Melanesia, economics are based on the so-called "big man" system, where a high-status individual controls resources and distributes them to the population through feasts to enhance his status.

Market economies are relatively impersonal. Although personal contact is involved at many levels, from commodities and stock trading to high-street retail, there is no necessity for any social bond between people who trade with each other, and much trade is between people who may not even meet. Moreover, the entire system is based on an impersonal standard currency with a value set by economic mechanisms rather than traditions.

As C rose over the centuries, the market economy became more central to economic life and more divorced from non-market forces. During the Middle Ages, prices were set and competition limited by social entities such as the guilds and boroughs. The market was more in evidence than among hunter-gatherer societies, but personal elements remained strong.

By the nineteenth century, as infant C reached an unprecedented peak, the personal social controls on the distribution and trade of goods had all but disappeared. Laissez-faire economics dominated Western Europe, and European trading empires encircled the globe. The guilds—or at least their power—had gone. Other medieval regulations, such as restrictions on usury, had also vanished.

Attitudes towards work

Animals in food-restricted environments spend a great deal of time searching for food even when not hungry, something also seen in the exploratory behavior of food-restricted rats. In the same way, humans with high C have a natural inclination for routine work and are more willing to forego socializing and leisure activities to perform it. The economic necessity of prestige may still drive them, but working hard is *easier*.

People in the Middle Ages were far less industrious than their descendants. Mondays were frequently treated as holidays, and the calendar was full of saints' days on which work was forbidden. In medieval England, roughly a third of the year was taken up by holidays. Southern European countries were even more relaxed, with holidays eating up about five months of the year.[126]

By the nineteenth century, working hours had increased enormously, and holidays were drastically reduced.[127] In 1840 the number of hours worked per year by an average manufacturing worker was between 3,105 and 3,588 (more than 10 hours per day, 6 days per week). By way of comparison, in 1987 the average number of hours worked in UK manufacturing had fallen to barely half that.[128]

The Victorians were not only willing to work hard but more willing to save for the future and invest in skills and knowledge that had no immediate benefit but which would pay off in the future. For example, literacy rose dramatically between the sixteenth and nineteenth centuries, well before the rise in wages generated by the Industrial Revolution.[129]

Charity and practical morality

One of the characteristics of high C is that it causes people to behave in ways determined by rules and customs, rather than personal sympathies. This applies as much to morality as to laws and government. In the matter of charity in particular, there was a dramatic change in approach between the Middle Ages and the Victorian era.

In the Middle Ages, caring for the poor was seen as a mandatory and virtuous duty of the clergy. The moral teachings of the church were largely concerned with correct ritual, but service to the poor was regarded as an unassailable good. Following the example of St Benedict, clergy were expected to receive and be charitable to poor strangers just as they would to known visitors.

However, in the thirteenth century the notion of the "respectable" poor began to be emphasized, and as the centuries passed, the requirements of respectability, and therefore the degree to which one deserved charity, increased. For the Victorians, the moral requirements for receipt of charity were severe. The deserving poor had to attend church, stop drinking alcohol, and abstain from inappropriate sexual activity. Lengthy, exhausting interviews were conducted to ascertain the moral worth of a potential recipient. The content of religion also changed, with a far greater emphasis placed on the virtues of self-help, character and moral improvement. Giving freely to strangers was no longer the way to divine approval.

The workhouses were the ultimate embodiment of the Victorian attitude to the poor. Grey, depressing institutions, they hammered home the message that any form of work was better than living off the state. The workhouse fostered an ethos of moral rehabilitation. Food was plain, life was monotonous, and all reading material other than the Bible was banned, as were dice and alcohol.[130] Prayers were twice daily and there was little time off.[131] Most controversial of all was the separation of families when they entered the workhouse, a powerful deterrent to seeking state assistance. Men could not apply alone but had to bring their entire family, who were then separated into dorms and exercise rooms.[132] The Victorian belief in the value of work and moral rehabilitation was so strong that even when inmates had no useful work to do they were assigned meaningless tasks to keep them busy. In 1884 Great Yarmouth Union workhouse took action to prevent inmates from loafing about:

> [The inmates] were put to work in a shed divided up into fourteen cubicles, five and a half feet square (one man to each cubicle, which was locked) shifting shingle with a shovel through a hole ten inches square. They had to move a ton an hour. They worked a nine-hour-day with one hour allowed for lunch. The shingle was then carted back to its place of origin.[133]

It is worth noting that all these requirements—the relentless focus on work, the restriction of sexual activity, the plain and limited food, the role of religion—are C-promoters. They reflect both the rise in C and the behavior that drove it to such a high level.

Science, technology and innovation

Rising C between the Middle Ages and the nineteenth century changed family and childrearing patterns, delayed marriage and puberty, and altered attitudes to sex and morality, all of which had a profound impact on the political and economic systems of Western Europe. But one key effect of C, and especially infant C, was to have a profound impact on the world. This is that huge surge in innovation and technology known as the Industrial Revolution.

An aptitude for machinery is one of the key characteristics of high infant C. The Manus people of the Admiralty Island, though stone-age tribesmen in dugout canoes, were uncannily skilled with machines when given access to them. The Japanese of the mid-nineteenth century, though locked in a feudal political system, showed an uncanny grasp of technology when shown the way. Both of these societies, like Europeans, began controlling children at a very young age. And both were like Europeans in being unusually open to change.

From Isaac Newton's scientific advances to the revolutionary new steam engine of James Watt, the advances in science, technology and the applied arts that took place between the seventeenth and nineteenth centuries were without parallel in human history. The new theories of gravity, bacterial infection, the steam engine, the steel constructions of civil engineering and of the new maritime age—all these things remade the Western world.

The early Victorians were so enamoured by machines that much of their patriotic pride was centred on the technological and industrial achievement of the nation. To them, the machine was an emblem of man's conquest of the forces of nature through the power of the intellect.

If major innovations in science and technology between 1455 and 2004 are plotted graphically against population size, as in Fig. 6.2 below, we can see the steepness of the rise.[134] This kind of assessment is bound to be subjective, but the pattern shown by the graph closely matches lists drawn up by a variety of researchers.[135] All show a rise in innovations to a peak in the nineteenth century.

Fig. 6.2. Innovations in science and technology relative to population size.[136] Based on 8,583 key scientific and technological innovations, plotted against population.

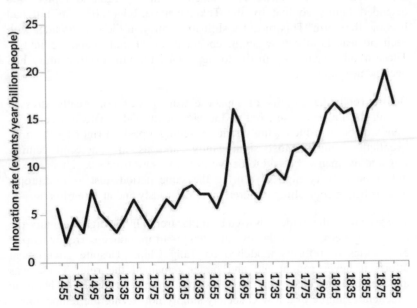

Japan

We have used nineteenth century Japan as an example of a society with relatively high infant C (if not at northern European levels) as an explanation for its rapid rise to industrial power. A study of Japanese history shows the same pattern of rising C as in Europe over roughly the same period. This is all the more interesting because Japan had no contact with the West before the sixteenth century, and was very little influenced by it before the nineteenth. Yet the trends in both societies had a great deal in common including changes in sexual and reproductive behavior, family structure, economics, education, and industry.

In twelfth-century Japan polygamy was widespread. It was common to take a secondary wife if the first did not produce an heir. But from the fifteenth century onwards this practice declined and was replaced by adoption. Wealthy men without sons would adopt a promising young man and make him his heir, marrying him to one of his daughters, known as the practice of *mukoyoshi* ("adopted son-in-law").

The extended family was powerful in early Japan, with local sub-branches providing labor to the main house in return for economic security. From the fifteenth century onward the nuclear family began to replace the extended family so that by the Tokugawa period (1603–1867) it had become the norm. This meant a shift to tenancy in the countryside, with each nuclear family now an independent social and economic unit.[137] These trends began first in the towns but by the nineteenth century had become near universal. [138]

Before 1600, the chastity of Japanese women was not greatly stressed among the lower classes, but in the eighteenth and nineteenth centuries standards became increasingly strict, reaching a peak of rigor by the mid-twentieth century. Most significantly, and as in nineteenth-century England, women were said to derive little pleasure from sex. This is a key indication of very high C. During the same period, just as in Europe, parents began controlling children more rigorously and at an earlier age.[139]

The growth of the market showed a similar trend. In the eighth century the imperial government made several attempts to introduce currency as part of a series of reforms modelled on Tang China. Despite government efforts it never caught on, and barter remained the rule. Trade was undeveloped and there were no cities apart from the capital.[140] This is an excellent example of a low C society simply not having the temperament for trade or the use of money, no matter what the government might decree.

Instead, people spontaneously shifted from barter to cash between the twelfth and fifteenth centuries, using imported Chinese coins because there were no longer any government mints. This period also witnessed the rise of professional moneylenders, commercial cities and vigorous overseas trade. Commercial growth continued during the Tokugawa era, with a shift from subsistence to cash crops and the growth of local industries such as cotton spinning and weaving. Even farmers and members of the lower class became increasingly literate, just as was happening in Europe at this time.[141]

Japanese history, like that of Europe, shows a steady rise in the indications of C—monogamy, the nuclear family, sexual restraint and the market economy. By the time Admiral Perry sailed into Tokyo Harbor in July 1853 with a demand that Japan open up to foreign trade, the C (and especially infant C) of the Japanese was such that they could readily accept and make use of Western science and technology. Within only a few decades, Japan had rocketed into the modern industrial age.

This success was in total contrast to the reforms of the seventh and eighth centuries, when the government had tried to turn Japan into a "modern" society like that of Tang China. As mentioned, attempts to introduce coinage failed completely, as did efforts to set up local administrations on Chinese models.[142] But one aspect of the reform was an unqualified success. Chinese Buddhism, adopted first by the court, came to dominate Japanese life and culture. Assisted by the later importation of Neo-Confucianism from China, its tenets drove the increase in C over the centuries that followed, playing the same role in Japan as Christianity in Europe. Not only did high infant C make fast industrialization possible after opening up to the West, it allowed an astonishingly fast recovery from the devastation of the Pacific War. This is the true basis of the "Japanese economic miracle."

Summary

What took place in Europe and Japan between the medieval and modern periods was a fundamental change in the temperament of the population. In Europe C-promoting behaviors, largely in the form of religious observances and sexual restraint, escaped from the monastic environments in which they had been nurtured and worked their way into the population at large. The result was a rise in the general level of C. It came slowly at first, but accelerated with the passing of the centuries, driven on by the effect feedback cycle. Control of children, restrictions on sexual behavior and personal discipline drove the rise of the nuclear monogamous family, industriousness, market orientation and innovations in science and technology. The wealth and power of the West, including both European descended peoples and Japan, are direct products of this change.

But there is more to the story than this. The swift rise of C in both Europe and Japan was driven not just by C-promoting cultural systems but by changes in the level of V and stress. The interaction of C, V and stress forms a cyclical pattern which explains not only how civilizations arise but also why they fall. This "civilization cycle" is the subject of the next chapter.

Testing

If the epigenetic signature of infant C could be identified by a test of blood, urine or saliva, as indicated in the previous chapter, more prosperous countries would be expected to have higher levels.

CHAPTER SEVEN

THE CIVILIZATION CYCLE

Spare the rod and spoil the child.
—Proverb

Between the Middle Ages and the nineteenth century, there was a huge rise in the level of C and especially infant C in the West, reaching an unprecedented peak in the Victorian period. In Japan a similar rise occurred over the same period but peaked in the early-twentieth century. In both nations rising C caused dramatic advances in industry, science and the market economy. Changes in religious and cultural practices promoted changes in temperament which built industrial civilization.

These changes had wide-ranging effects, not confined to the nations concerned. Societies whose cultural practices promote C and V—through control and punishment of children, sexual restraint, control of women, and so on—acquire political, economic and military power which allows them to dominate societies lacking these traits. And with a high level of V, they have not only the means but the *will* to expand and dominate. The European colonization of North America is one example.

Thus we see the importance of human temperament and the physiological systems that underpin it. But if we adjust our focus and examine the development of a civilization over a longer timescale, we find that changes in the levels of C (and indeed V) do not follow a simple linear path. Instead, the rise of C is accompanied at first by a rise in V and stress, and then by a fall. And as C falls, the levels of V and stress first continue to fall, and then begin to rise. This is the Civilization Cycle, and in this chapter we consider how it works, at least for the rising C section of the cycle.

C-promoters are not enough

From the Middle Ages onward, Christianity was instrumental in driving the rise in C in Europe. It contains powerful C-promoters such as sexual abstinence, self-discipline and self-denial, fasting, and ritual observances.

Given this basic principle, we ought to expect a high level of C in any civilization where Christianity is firmly entrenched. In fact, we do not. Christianity grew steadily in the Roman Empire until it became the official religion. And yet, as we shall see in a later chapter, C actually *declined* during the late Empire, continuing a trend which had been going on since the mid Republic. Eventually, the decline reached the point where the Western Empire fell apart. Not only did Christianity fail to raise C in the Roman Empire, it could not even halt the decline. It was not until long after Rome adopted Christianity, and several centuries after its spread throughout Europe, that C began to rise towards its nineteenth century peak.

If Christianity did not cause C to rise in the first millennium AD then it cannot wholly explain its rise in the second millennium. There must be another factor missing from our model, some other element which is required in addition to the C-promoters we know about.

The missing factor is V, which is also reflected in the level of stress.

V, Stress and the rise of C

In this chapter we will see evidence that in both England and Japan, the long rise in C to the nineteenth and early twentieth centuries (respectively) was accompanied by a distinct rise-and-fall pattern in stress. Indications include all the behaviors seen in highly stressed Whipsnade baboons— extreme status differences, brutal authority, harsh punishments, suspicion and fear. Social hierarchies were also highly unstable. The proposed reason is that there was also a rise and then fall in the level of V, as evidenced by the fact that patriarchy follows much the same pattern. Levels of stress tend to mirror levels of V because high V reduces tolerance for population density, as indicated in chapter four. Thus, all things being equal, rising V must increase the level of stress.

The civilization cycle in English history

Patriarchy

The most direct evidence for a peak in V in the sixteenth century can be found in attitudes towards women, remembering that patriarchy is one of the key indications of high V. In the thirteenth and fourteenth centuries, women were increasingly downgraded both legally and theologically. In 1317, for instance, they were barred from the throne. By the sixteenth century, their legal entitlements had declined to such a state that they required their husband's consent for any legal act to be valid.[143]

In the early seventeenth century, however, preachers began to stress companionship in marriage, linked to a general sense of greater equality. The rise in woman's status would continue into the nineteenth and early twentieth centuries when country after country gave them the right to vote.[144] Some people might consider control of a woman's sexuality as inseparable from her subordination, but in fact they have quite different effects. Low status for women is a V-promoter, as it causes anxiety which is transmitted to infants. Control of women's sexuality is a V-promoter but also a C-promoter. Thus, it is quite possible for women's status to improve while sexual mores tighten, as was the case between the seventeenth and nineteenth centuries.

Punishment of children

More indirect evidence for a peak of V can be found in the level of stress. One indicator is the practice of punishing children, which peaked in and around the sixteenth century in England.

During the Middle Ages, particularly in the twelfth and thirteenth centuries, authorities on social ethics advised against the beating of children. But from the fifteenth century onward, that changed radically, with a pronounced trend toward greater punishment.[145] Flogging became more prevalent, reaching an extraordinary level of ferocity in the sixteenth and seventeenth centuries.

In 1612, the puritan schoolmaster John Brinsley elaborated on the bible passage "spare the rod, spoil the child" in his dialog on education, *Ludus Literarius*:

> Finally, as God hath sanctified the rod and correction, to cure the evils of
> their conditions, to drive out that folly which is bound up in their hearts, to
> save their soules from hell, and to give them wisdome, so it is to be used
> as God's instrument to these purposes. To spare them in the cases is to
> hate them. To love them is to correct them betime. Do it under God, and
> for him to these ends and with these cautions, and you shall never hurt
> them: you have the Lord for your warrant. Correction in such manner, for
> stubbornnesse, negligence and carelessnesse, is not to be accounted over-
> great severitie, much lesse crueltie.[146]

Schoolmasters were enjoined to exercise restraint and dignity, and avoid
anger or savagery when whipping children, but the punishments
recommended seem positively ferocious to modern values. Also, the
cautions to restraint recommended by writers like Brinsley were not
necessarily followed in practice. Children could be given up to fifty
strokes with an elm rod or birch simply as a reinforcement to rote learning.
Pulling by the ears, and lashing across the face and head (which Brinsley
specifically advised against) were common. In 1582, at the age of 11,
William Bedell (later a churchman and martyr of the Reformation) was
attacked by an enraged schoolmaster:

> [He was] knocked down a flight of stairs and hit so violently … that blood
> gushed out of his ear, and his hearing was in consequence so impaired that
> he became in process of time wholly deaf on that side.[147]

Such severe punishment of children was generally accepted, and it was
rare for such masters to be dismissed.[148] It was so commonplace that
beatings were not confined to schoolchildren. Even undergraduates were
flogged until 1660.

But from around this time a reverse trend can be observed. Parents began
to be more lenient during the seventeenth century, and this trend continued
through the eighteenth, especially among the middle classes. Beating of
schoolchildren remained ubiquitous but became gradually less severe
during the eighteenth century, and brutal schoolmasters were more likely
to be publicly condemned.

In the early nineteenth century new ideas about childrearing emphasized
the importance of control but warned against physical punishment. The
writer and philanthropist Hannah More advocated tighter control of
children, but using psychological means rather than physical punishment.

Judicial Punishments

If children were treated with such severity, it can hardly be surprising that criminals suffered even worse. Torture, which had not been a part of judicial processes prior to the thirteenth century, became routine thereafter. Torture was used especially in the trials of religious heretics. With Christianity gaining an ever stronger influence in England, intolerance of heresy escalated in the late fourteenth century, marked by the passing of the Heresy Act of 1382.

As well as torture as an element in trials, and execution as a punishment, public whipping became increasingly widespread in the sixteenth century, even for quite minor offenses. A law was passed in 1531 under which unlicensed beggars were to be "stripped down to the waist and whipped," while the punishment for vagabonds was to be "tied to the end of a cart naked and be beaten with whips throughout the market town or other place until bloody."[149]

Violent punishments that increased in this period included dragging from a boat (which could result in drowning),[150] branding, racking, impalement, burning and boiling alive. Hanging, drawing and quartering was an especially nasty form of execution in which the living victim was half strangled and then castrated and disembowelled. First introduced in 1241, it drew large crowds. Indeed, all public punishments were well attended, reflecting the popular acceptance of such violence.

One gruesome punishment was the burning alive of religious heretics, especially associated in England with Mary Tudor (1553–1558), but also ordered by the otherwise humane and enlightened Thomas More, celebrated in the film *A Man for All Seasons* (1966). The following account is of the burning of Bishop John Hooper in 1555, after two earlier attempts had failed for lack of enough dry wood.

> After the second fire was spent, he wiped both his eyes with his hands, and beholding the people, he said with an indifferent, loud voice, "For God's love, good people, let me have more fire!" and all this while his nether parts did burn; but the fagots were so few that the flame only singed his upper parts.
>
> The third fire was kindled within a while after, which was more extreme than the other two. In this fire he prayed with a loud voice, "Lord Jesus, have mercy upon me! Lord Jesus receive my spirit!" And these were the last words he was heard to utter. But when he was black in the mouth, and

his tongue so swollen that he could not speak, yet his lips went until they
were shrunk to the gums: and he knocked his breast with his hands until
one of his arms fell off, and then knocked still with the other, while the fat,
water, and blood dropped out at his fingers' ends, until by renewing the
fire, his strength was gone, and his hand clave fast in knocking to the iron
upon his breast. Then immediately bowing forwards, he yielded up his
spirit.[151]

During the seventeenth century, the acceptance of violent punishment
gradually lessened. Torture became less widely used, the last recorded
cases being in the 1650s; though "pressing" with weights (as a punishment
for refusing to plead at one's trial) continued until the eighteenth
century,[152] but that too eventually died out. Capital punishment also
declined. In London and Middlesex between 1609 and 1799, the average
annual number of executions fell from 140 to just 33, despite a significant
increase in population. By 1830, the number of people executed
throughout the whole of England had fallen to 51.[153]

Formality and marked status differences

One of the characteristics found in highly stressed societies is an increase
in status differences and an acceptance of powerful authority. These were
all factors reflecting a rise in V and stress from the late Middle Ages to the
sixteenth century.

Class distinctions increased, as did the formality of social relations. Feasts
in noble houses embodied both features. Guests came to be seated
according to a strict plan determined by social rank, including gaps
between the different grades. Feasts were marked by formal ceremony,
and even conversation was bound by rules (or even forbidden altogether).
At one dinner attended by Elizabeth Woodville, queen consort to King
Edward IV, the entire court was required to kneel in silence while she
ate.[154] The donning and doffing of hats became a pervasive feature of
social life, the removing of a hat being seen as a sign of lower status.
Disputes over precedence were common and often resulted in violence,
such as gentlemen fighting over the right to "hold the wall," a favored
position when the contents of chamber pots were commonly hurled from
upstairs windows.

The relationship between generations within a family grew ever more
deferential. A man, regardless of his age, was expected to kneel to receive
his father's blessing, and to stand in his father's presence until invited to

sit.[155] This formality had a very different meaning from the strict social etiquette of the eighteenth and nineteenth centuries, which had more to do with courtesy than deference and is an indication of C rather than stress.

The increase in the scale and importance of social differentiation affected the practice of religion, and the role of Christianity in English society. In the Middle Ages, clergymen had felt quite free to criticize the behavior of the rich and the noble, and political conflict between church and crown was a constant feature of the English state. But by the reign of Elizabeth I in the late-sixteenth century, preachers taught humility and obedience to the crown.

Returning again to the subject of punishment, there was a sharp distinction between classes. During the trial of Anne Boleyn the musician Mark Smeaton, a commoner alleged to have committed adultery with the queen, was tortured brutally, whereas aristocrats found guilty of complicity, such as Henry Norris and Anne's brother George, were merely executed.

High V people under stress are willing to accept, and are more likely to obey, arbitrary and even brutal authority. Thus the rise of V and stress in the period leading up to the sixteenth century gave the Tudor monarchs a power not seen before or since in England. Earlier kings had been constrained by their barons, who had considerable power through the local support they enjoyed from the populace. But Tudor kings and queens enjoyed almost total power, dictating religious policy and executing ministers and consorts with little regard for law or justice.

Respect for hierarchy and deference began to ebb in the early seventeenth century. The Puritans were particularly active in this respect, and egalitarian political movements such as the Levellers gained significant traction as the country moved towards Civil War in the 1630s and 1640s. Parliament came into increasing conflict with the monarch, refusing to tolerate the kind of autocratic rule that had been the norm a century before.

By the eighteenth century, social formality had reduced. The aristocracy mixed quite freely with wealthy commoners. Children, though increasingly controlled in their behavior, were no longer required to show the same ritualized formal deference to their parents as in previous generations, and formal deference to one's parents was no longer required from adult offspring. There was some retention of formality in rural areas, which maintained higher V and stress for longer, but even there it had declined by 1900.[156]

As V and stress continued to fall in the nineteenth century, Britain and Europe became gradually more democratic. In Britain, the Reform Act of 1832 marked a major change toward democratic rule and popular accountability. From that point on, it was only a matter of decades before all men were able to vote. In France universal male suffrage came in 1848, and in Germany after 1871. People had lower V and stress than their forebears, and were less willing to accept the power of kings and aristocrats.

Unstable hierarchies: conspiracies and treason

It is characteristic of societies under stress that status hierarchies are not only steeper but more unstable, as seen in the fate of English monarchs between the fifteenth and seventeenth centuries. From the Norman Conquest until 1327, English kings had suffered setbacks but had never been deposed or killed. In that year, however, Edward II was unseated by his wife and her lover, then imprisoned and murdered.

From that point on, the throne became increasingly unstable. Richard II was dethroned by Henry Bolingbroke in 1399. Henry's grandson Henry VI was dethroned twice and then put to death in 1471. His successor Edward IV was briefly ousted in 1470. Edward's son was deposed in 1483 and then murdered along with his brother. Finally, Richard III was defeated and killed at the battle of Bosworth Field in 1485.

In the next century and a half no monarch was deposed or killed, but conspiracies were a feature of the age. For instance, John Dudley took control of the government during the reign of Edward VI, and attempted to change the succession when Edward died in 1553 by placing Lady Jane Grey on the throne. He failed and was executed, as was she by the following year. Elizabeth I (1558-1603) faced constant threats of assassination, foiled only by the efforts of her brutal and efficient spymaster, Sir Francis Walsingham. It was one such conspiracy that led her to execute her cousin, Mary Queen of Scots.

When James I came to the throne in 1603 he immediately faced two major conspiracies, the Bye plot and the May plot. Then, in 1605 came the Gunpowder Plot, in which Guido Fawkes and his co-conspirators attempted to blow up both the king and parliament. James' son Charles I was deposed and executed in 1649, and his grandson James II was deposed and exiled in 1689.

After 1689 politics also became much more stable. No monarch or prime minister in Britain was again to be overthrown by violence (unless one counts the assassination of Spencer Perceval in 1812, which was a personal grudge rather than a political coup) and certainly none were executed. Personal relationships in general also became more stable. Concepts of manliness in the eighteenth century became less dependent on bearing arms or committing violence if offended.[157] There was also less violence at all levels of society.

Bad Temper and suspicion

People under stress tend to be suspicious of each other, to expect evil of others and to plot against them, one result being political instability as described in the previous section. Suspicion is a behavior already observed in the highly stressed Mundugumor, and was a characteristic of Englishmen in and around the sixteenth century. In 1485 it was said of Englishmen that they were "reckless and ruthless, guided by temper, with higher ranks arrogant and all prone to expect evil of their fellows."[158] The population of early seventeenth century London has been described as volatile in the extreme, with people showing fear, grief or anger at the slightest cause.[159] Advice given by fathers to their sons – evidenced in letters written between the fifteenth and seventeenth centuries – indicates a cynical and pessimistic view of human nature. Young men were counselled to be secretive and mistrustful even within their families.[160] The Tudor era was also a time in English and European history when witch-hunts were at their absolute height. Hundreds of innocent women were tried and burned to death under the suspicion of Satanism.

By the nineteenth century there were increasing signs of calm, far removed from the passionate hysteria of the sixteenth century. An entirely new concept of masculinity was emerging to replace the prickly 'man of honor': the 'man of dignity' who was reasonable, prudent and with command over himself.[161]

In the nineteenth century, Englishmen acquired a reputation for sang froid and a "stiff upper lip." The new standards of restraint applied to all men, and there was a particular emphasis on treating women with deference.[162] The socially disruptive behavior of the sixteenth century had been replaced with the sober and controlled conduct of the Victorian gentleman, reflecting an enormous overall reduction in stress.

Thus it is that stress, like V, reached a peak in the sixteenth century and then declined, falling slowly during the seventeenth century and then faster in the eighteenth through to the nineteenth centuries, as indicated in Fig. 7.1.

Fig. 7.1. The civilization cycle in England. A high level of V and stress around 1550 was accompanied by a rapid rise in C. As V and stress fell to the nineteenth century, C reached a peak and then began to fall.

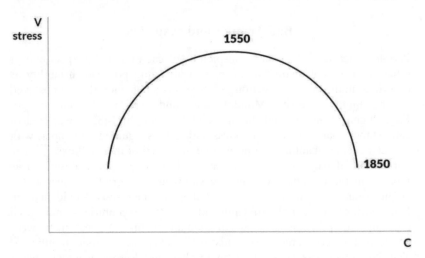

The civilization cycle in Japanese history

The histories of England and Europe between the Middle Ages and the nineteenth century demonstrate a very striking pattern. A rise in V and stress accompanies a rise in C. V and stress reach a peak level, at which time C is rising rapidly. From this point V and stress begin to fall, although C continues to rise for several more centuries.

If this is a physiologically determined phenomenon, we should expect to see it in other civilizations experiencing a rapid and sustained rise in C. The history of Japan during approximately the same period provides us with a case in point. Being so far removed from Europe, it makes a particularly good independent case study.

As mentioned in an earlier chapter, C rose dramatically in Japan in the nineteenth century, leading to a sudden flowering of industrial and economic modernization. It lagged somewhat behind Europe, but the processes leading up to it played out over approximately the same period.

Despite this lag at the highest-C end of the cycle (that's the rightmost point on the graph, where C peaks and begins to fall), the peak of V and stress occurred in Japan at almost the same time as in Europe – the late fifteenth or sixteenth century. This must be coincidence, as there cannot have been significant direct influence between the two cultures. The matching of the V peak and the mismatch of the C peak is our first indication that, if the civilization cycle is a universal pattern, it doesn't necessarily play out over a consistent timescale.

During the late first millennium AD, Japan seems to have been relatively low in V and stress. Imperial law codes in the eighth century AD were relatively gentle. Japanese culture was heavily influenced by Indian Buddhism and there was a reluctance to impose severe physical punishment.

Another indication of low stress is that Japanese rulers and social hierarchies were relatively secure. The Emperor was the nominal ruler but in practice a single family – the Fujiwara – held power until the eleventh century. During this time women enjoyed a comparatively high status, a common feature in societies with low V. They controlled their households and were generally influential, especially among the lower classes.

All this began to change after 1000 AD as the provincial military aristocracy gained power, and in the late twelfth century took over the state. Japan began to develop the aggressive warrior culture for which it is popularly known, a sign of rising V.

An indication of rising V and stress is that punishments became more severe. A new legal code introduced the death penalty for all manner of offenses, including theft. For the time being, the ruling hierarchy was stable, with the Hojo family until 1333 with few successful rebellions. That changed as V and stress continued to rise. During the Ashikaga period (1336-1573) brutality, strife and instability increased. Warfare became a way of life, and Japan's unified central authority collapsed most notably in the Onin War that broke out in 1467 There was mass slaughter in the streets of Kyoto involving marauding peasants as much as professional soldiers.

Like Europe, Japan at this time was also riven by suspicion and mistrust. Feudal lords were mistrustful of their subjects and kept a close eye on them, dictating to an extreme degree what they could and could not do. Outsiders were also kept under close scrutiny, or simply not permitted to enter fiefs at all.[163]

In the late fifteenth and sixteenth centuries local samurai established new and brutal law codes for their realms. Entire villages might be punished for the failure of one man to pay taxes. Methods of execution involved burning alive, boiling in oil, impalement, sawing, and crucifixion. Where previously only direct vassals of the Shogun were executed for rebellion, now even low ranking prisoners were tortured to death. European visitors were shocked by the brutality, despite coming from a culture which (as we have seen) was harsh and violent by modern standards.

One reason for this extreme harshness was the instability of the political hierarchy, brought on by chronic stress throughout the entire society. Punishments were an attempt to hold on to power. In the century leading up to the 1560s all but one of the great ruling families were either destroyed or reduced to insignificance, often by rebelling vassals. A number of foreign visitors noted the prevalence of treachery during this period.

Consistent with these signs of rising V and stress, the status of women fell markedly to the sixteenth century, to the point where the power of men became theoretically absolute. Patriarchy was so strong that men were not supposed to consult their wives even over family matters.

Just as in Europe, many of these trends began to reverse in the seventeenth century, following Japan's unification. Out of the constant upheaval emerged a single ruling lineage, the Tokugawa dynasty, which reigned for 265 years from 1603 to 1868. Central authority was largely stable and accepted, and there were few attempts to overthrow the state. Stability within the state also improved, and senior ministers were relatively safe from suspicion. Many served as shogun and retired peacefully.

Children also began to be treated less harshly. In the fifteenth century it was assumed that they would be beaten, but by the eighteenth century parents were warned against anger when dealing with children.[164] Physical punishment continued to decline to the point that it became rare in Japan by the 20th century.

So it seems that the civilization cycle model fits Japanese as well as European history. One difference between the two, however, is that the evidence for falling V in the Tokugawa era is not as clear as it is in Europe, perhaps because a conservative regime may have slowed its decline. The status of women in Japan remained low. Punishments during the Tokugawa period were also still severe by Western standards, if not severe as they had been. Only during the Meiji restoration in the 1870s, which began the wholesale modernization of Japan, were legal limits placed on the number of crimes punishable by death. Legal codes in 1925 became more lenient still, a trend which has continued up to the present day.

Japan, like England and Europe, experienced a sharp rise in V and stress to the late fifteenth and sixteenth centuries, but the decline seems to have been slower than in Europe, which may help to explain why C did not peak until a century later, as indicated in Fig. 7.2.

Fig. 7.2. The civilization cycle in Japan. As in England, a high level of stress and presumably V is associated with a rapid rise in C, and high C with falling stress and V.

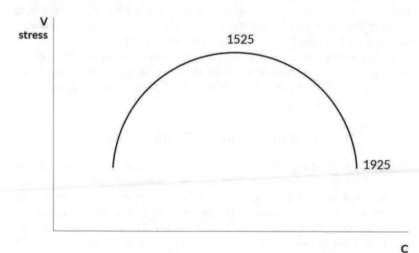

We now have a picture in which a rapid rise in C in both England and Japan was accompanied by a peak of V and stress around the sixteenth century. As mentioned earlier, the high level of stress can be explained as a direct response to high V. As detailed in chapter 4, V makes people and animals intolerant of crowding, which is a key part of the biological function of V. By reflecting past experience of famine it impels populations to migrate before population density becomes overly high, thus making it more likely they will find a refuge when the next disaster strikes. In a dense agricultural population where people cannot migrate, a rise in V causes an increase in stress.

The reason for the rise in C is a little more complex. In both cases, during early periods when V and stress were at only moderate levels, C does not appear to have risen at all. When V and stress were very high, C rose rapidly, and then, once they had dropped back to a moderate level C ceased to rise altogether and began (as we will see) to fall.

This suggests that in both Japan and Europe a major increase in C was apparently driven by a surge in V and stress. One obvious link is through the stress hormone cortisol. The harsh childhood punishments that raise V and stress involve a cortisol reaction which toughens the stress reaction. Stress itself involves elevated cortisol. We saw in chapter 2 that cortisol is a C-promoter in that it inhibits testosterone. Another is that the rigorous social regime required to effectively enforce C-promoters needs a hierarchical and highly deferential culture to work effectively. People in cultures that have high V and stress are likely to obey religious authority, which would in turn be more likely to punish dissent. Thus, what we are seeing is a surge of cortisol around the sixteenth century in both Europe and Japan which, in conjunction with the increasingly strict C-promoting traditions in Christianity, Buddhism and Confucianism, acted as a powerful C-promoter.

What drives the civilization cycle?

The final question regarding this cycle is why V rose so high in the first place. The key for Europe can be found in Christianity, which as we saw in chapter 4 is not only a C-promoter but a V-promoter. V-promoting aspects of Christianity include patriarchy, the requirement for sons to honor and obey their fathers, the annual Lenten fast, and sexual restraint. After growing in both strength and authority in Europe after the twelfth century, Christianity would have steadily increased the level of V in Europe. At the same time Buddhism and Confucianism played a similar role in Japan. The rising level of V in both Europe and Japan also coincided with frequent outbreaks of famine, which would also have supported higher V.

There is another factor behind the rise of V in the early Middle Ages. Recall from chapter 4 that V is highest when mothers are anxious but also highly indulgent. This means they do not punish or control their infants in any way. It follows that early control of children, which results in infant C, must undermine V.

Until about 1500 C and especially infant C were low, which means that infants were treated indulgently. This made it possible for V to rise, driven by Christianity and Buddhism and with a nudge from occasional famines. But as V and thus stress rose, C increased and (as we have seen) people began to control children earlier in childhood. This slowed the rise of V until it stabilized at a peak in the sixteenth century. From then on the continued rise in early childhood control began to undermine V, and as C rose still further the fall in V accelerated. By the nineteenth century C was at a peak but V was in steep decline. These relationships can be summarized as follows:

1. Low C permits V to rise, which in turn raises Stress.
2. High V and stress cause C to rise
3. As C and especially infant C rises, they cause V and thus stress to fall.
4. The fall in V and stress eventually causes C to fall.

In chapter 12, on the fall of Rome, we will see that when C falls to a certain level, V begins to rise and we return to step 1.

Thus it is that the dramatic rise of V and stress in Europe and Japan around the sixteenth century is far more than a mere historical curiosity. It represents a surge in cortisol which drove the rapid rise in C and thus made possible the overwhelming success of Western civilization. As will be discussed in future chapters, the civilization cycle and the mechanisms behind it also make clear why that civilization must fall.

Testing

Two groups of rats are subjected to mild food restriction but with different levels of stress. The more stressed animals should show a greater level of C effects.

CHAPTER EIGHT

LEMMING CYCLES

Thus the superior man
Understands the transitory
In the light of the eternity of the end.
—I Ching

The civilization cycle is not enough

We now have a good working model of the progress of a civilization over many centuries that applies to at least two civilizations. A surge in V and thus stress combine with powerful C- and V-forming traditions to drive a substantial rise in C. Between the Middle Ages and the nineteenth century, Europe and Japan passed through the "upper" half of the cycle where C is rising, while V and stress rise to a peak and then decline.

This model explains the increasing strength of government, the move from monarchical systems of government to more impersonal, democratic forms, the rise of market economies, and the rapid advancement of science and industry with the accelerating rise in the level of C.

But there are weak spots in the civilization cycle model—details that it does not explain. For instance, according to the cycle model, the governance of England in the mid-fifteenth century should have been growing stronger as C rose, bringing an increasing acceptance of authority. But instead there was an outbreak of feudal disorder in which several kings lost their thrones. A short while later Japan experienced a similar episode known as the Onin Wars. Chinese history contains a number of such periods of disorder, which typically cause ruling dynasties to collapse – despite a broad trend to more stable government over the past two thousand years.

In more recent history there was a massive outbreak of aggressive warfare in Europe in the early twentieth century, following almost a century of relative peace.

What can we make of these anomalies? The civilization cycle is clearly robust in so many ways but does not account for these awkward exceptions. Once again, as with previous problems encountered in the development of biohistory, demography supplies the clue.

In both Europe and Japan, until the nineteenth century there was a trend for faster population growth associated with the civilization cycle. But that broad trend masks variations in the *rate* of population growth. When examined within local populations a distinct pattern can be observed around three hundred years in length. This represents a cycle within the civilization cycle, which also correlates with other changes in behavior and attitude.

However, before examining the evidence we will return to the field of zoology, which will indicate that humans are not the only species to show these shorter-term patterns of population growth and decline.

The lemming cycle

The one thing most people "know" about lemmings is that they commit mass suicide by hurling themselves from cliffs. This is a complete myth, as we will see, but one based on a remarkable pattern of behavior.

Lemmings are small rodents living in the sub-arctic regions of Eurasia and North America. Every few years their populations expand enormously and their behavior also significantly changes. Normally timid and shy, at their peak of population lemmings lose their fear of humans. They become bold and even aggressive, spreading out in a wave of mass migration. The urge to migrate is so strong that groups of lemmings attempt to cross any obstacle standing in their path, including lakes and rivers. Large numbers drown attempting to cross lakes or seas—which is where the "suicide" myth originates—but some succeed in crossing quite large bodies of water.

This dramatic expansion and migration is followed by a population collapse. Then there is a new period of growth until a new peak and mass migration.[165] The cycle is very regular, playing out over a three- to four-year period.

Lemmings are not the only species to display this pattern. Mice, hares, grouse, muskrats and many others have similar population cycles, although typically lasting ten years rather than three to four. Indeed, one authority has claimed that the only non-cycling species he could identify in North America was the beaver.[166] Not all of these species exhibit the same dramatic, frantic migration as lemmings, but the essential pattern is the same.

Growth, migration and regularity are not the only remarkable features of the cycle. Muskrats have been studied intensively and show some clear differences in behavior. During the growth phase they are more resistant to disease, breed earlier and more profusely, and are more tolerant of population density. During the decline phase, the opposite occurs: more disease, smaller litters, and breeding sites further apart (see Fig. 8.1 below). They also seem to react poorly to any form of stress, commonly dying from shock if taken into captivity.[167]

Fig. 8.1. Vital statistics of Iowa muskrats. In the growth phase of the cycle litters are larger, disease resistance is stronger, and more animals breed in the year of their birth. All these factors work together to promote population growth.

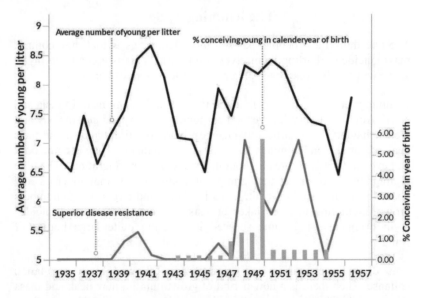

The length of the cycle seems to be determined to some extent by the average breeding age of a species. Lemmings, which can reach sexual maturity in just a few months, have a three- to four-year cycle. Muskrats, which usually only breed in the year following their birth, have a ten-year cycle. There is some evidence that very large, long-lived species, such as elk, have lemming-type cycles which run over much longer periods. Based on muskrats and similar species, such as varying hares, we suggest that these cycles last around ten generations.

Explanations of this pattern have centred on population density or food shortage as reasons for the migration and population collapse, but observation has shown that the cyclical pattern does not depend on actual population size. Even in local sub-populations that were quite low to begin with, the collapse still occurs. There is no generally accepted theory of lemming-type cycles—why they happen and what drives them along. There is not even a standard term for them, so in honor of the best-known cycling species we will simply refer to them as "lemming cycles."

One way to describe lemming cycles is as a roughly ten-year alternation between V and C. During the growth phase populations are bolder, migratory and breed more—all characteristics of V. During the decline phase they breed slower and become more timid, with larger territories, which are characteristics of C (see Table 8.1 below)

Table 8.1. Behavioral characteristics of lemming cycles. Peak populations have V characteristics and trough populations C characteristics.

Growth or high population phase (V)	Decline or low population phase (C)
Higher birth rate	Lower birth rate
Lower death-rate	Higher death-rate
Bolder	More timid
Less Territorial	More territorial
Migratory	Less migratory

Lemming cycles in human history

This notion of an alternation between V and C fits even better when we consider human societies, and explains many of the anomalies pointed out in earlier chapters. The important point is that lemming cycles reflect changes in V and C *within the overall long-term trend*. The civilization cycle involves major changes in C and V over many centuries. The lemming cycle involves much smaller rises and falls. So, for example, the long-term trend for C may be upwards if it is on that part of the civilization cycle, but at any one moment it might be rising *or* falling in the shorter term, depending on where the society is on the lemming cycle. It is as if the civilization cycle represents the tides and the lemming cycle the waves, with the actual height of the water depending on both.

Lemming cycles—the model

The lemming cycle can be identified by one simple and measurable factor—population growth. Therefore, we should be able to construct a basic predictive model that can be tested quite easily against the historical evidence. If the cycle truly represents fluctuations in C and V, the test will be whether humans behave accordingly at different stages of the cycle. Are there changes in V- and C-type traits at the appropriate moments, not just in basic mammalian traits listed in Table 8.1 but in specifically human attitudes such as political loyalties and resistance to change? An initial model is illustrated in Fig. 8.2 below.

There are two vital points on the cycle, the most important labelled "G" is the point of fastest growth (but not maximum population size, which occurs later). The opposite point on the cycle is the point of slowest population growth or fastest decline. Again, this does not necessarily mean that the population size is smaller than it was at the start of the cycle, or that it has reached its lowest level, only that the rate of decline has bottomed out.

As we know, V is associated with rapid population growth and warlike aggression (see chapter four). Therefore, we might suppose that the peak of V and the G point would go together. In fact, they do not. Migration (which is one expression of V) tends to happen when animals born at the peak of growth become adults a generation later. But this is consistent with the behavior of cycling animals, which tend to migrate *after* the time of fastest growth, when population size is at a maximum.

We should expect to see three separate peaks. The first is G, the peak of population growth. The second peak is in the level of V, and occurs a generation after the G point. The third peak, another generation later, is in child V, which is caused by the high level of V in the previous generation.

Fig. 8.2. Lemming cycle initial model. A generation after the peak of growth (G) is the peak of V, and a generation after that the peak of child V. A generation after the low point of growth is the peak of C, and a generation after that of infant C.

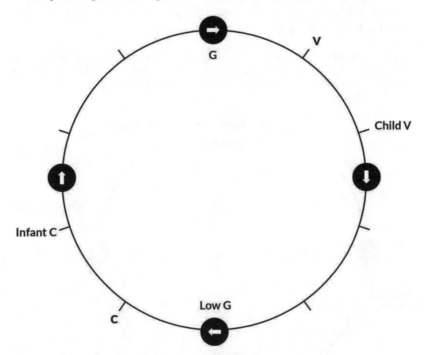

Likewise, we should expect a complementary series of critical points in the decline phase. The slowest population growth or most rapid decline is at the low point of G. Then a generation later comes the peak of C (assuming C and V are in direct alternation), and a generation after that the peak of infant C caused by maximum C in the preceding generation.

We can estimate the length of a full lemming cycle in a human population by taking the average length of a generation. This will vary widely with different populations, but since humans commonly marry in their late teens or early twenties and are less fertile after age forty, a typical human generation might be thirty years. In animal populations, the typical cycle length is ten generations, which suggests a human lemming cycle of three hundred years.

To mark our key points (G, V, child V, low G, C and infant C), the lemming cycle is divided into ten sections of thirty-year generations. Thus, V peaks at G+30 and child V at G+60. On the other side of the cycle, C peaks at 120 years before the G year (G-120), and infant C at thirty years after that (G-90).[168] Fig. 8.3 plots these points.

Fig. 8.3. Lemming cycle model. Assuming a three-hundred-year cycle with thirty-year generations, the peaks of V, Child V, C and infant C can be plotted as a number of years before or after G, the time of maximum growth.

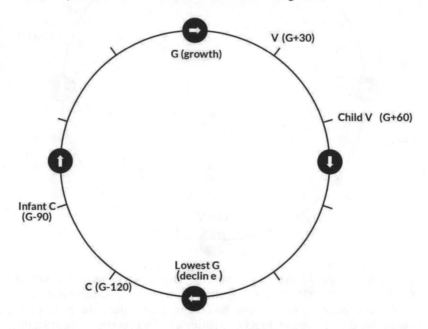

With these markers in place, our initial model makes predictions about the temperament and behavior to expect at various key points on the cycle.

G represents the peak of population growth and gives rise to V in the next generation.

G+30 is the peak of V, a trait we understand quite well. We should expect a peak in confidence, aggression and expansion at this point.

G+60 is the peak of child V and the low point of infant C. We should see weak local loyalties and acceptance of powerful authority, with more traditional attitudes.

G-150 is the low point of G. Birth rates should be lower and death rates higher.

G-120 is the high point of C and the low point of V. People here should be economically skilled, breed late and be least likely to engage in wars of economic expansion.

G-90 is the peak of infant C and low point of child V. This means that loyalties should be intense and local, with a strong preference for leaders who are similar in culture and language. People should be less obedient to distant authority, and local aristocrats more powerful. There should also be an openness to change.

Testing these predications against actual social and political changes in England over the past thousand years, what we find is not completely in line with these predications. However, the pattern is consistent and in many ways remarkably close.

Lemming cycles in England

Patterns of population rise and decline are fairly well known through much of the period in question, although exact figures are only available from the nineteenth century. Before that, the numbers are approximate but sufficient for our purpose.

From the late eleventh century, the population of England was on the rise. The rate increased in the twelfth and thirteenth centuries. Then, in the late fourteenth and early fifteenth centuries it declined. Another rise began in the late fifteenth century, continuing into the seventeenth century. After that, population remained static in the early eighteenth century before rising in the late eighteenth to early twentieth centuries. This gives three surges in and around the thirteenth, sixteenth and nineteenth centuries (see Fig. 8.4 below), consisted with a predicted lemming cycle period of three hundred years. G years can be set around 1230, 1550, and 1850.

It is important to note that these are not necessarily the years when population was growing fastest. For example, in the nineteenth century this actually happened in the 1810s and the 1870s, with a dip mid-century the result of a third cycle, to be discussed in chapter ten. Lemming cycles are a major factor affecting population growth, but not the only one.

Fig. 8.4. Population growth and decline in England, 1000–1900. Growth takes place in three distinct surges, separated by periods of static or declining population. This cuts across the trend to higher population growth associated with the long-term rise in C.

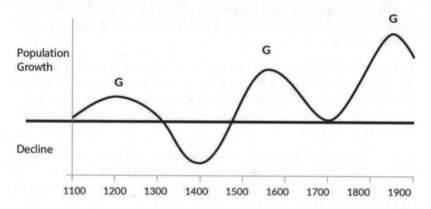

The G-150 years, which show the slowest growth or greatest decline, are 1400 and 1700. These are roughly mid-way between the periods of fastest growth, which is what we would expect from a lemming cycle pattern.

This gives two complete cycles—1230–1550 and 1550–1850. Note that the second cycle is twenty years shorter than the first, something to be considered later.

As with muskrats, the slower growth in the "trough" periods seems to have been a result of fewer births *and* more deaths. Muskrats in this phase were less resistant to disease, and so were humans, historically. The Black Death, which killed around a third of the population of England in 1348 and many more during the next 150 years, might be thought to account for the decline prior to 1400 shown in Fig. 8.4 above. But the plague, while devastating, does not explain the broader demographic patterns occurring at the time. The English population had in fact been static or falling since at least 1300, well before the plague struck.

Moreover, evidence from a plague cemetery in East Smithfield, London, shows that victims were more likely than the general population to show evidence of malnutrition or other weaknesses in the bones.[169] This suggests that the plague claimed those who were already weak, and may not have lived long in any case. In addition, plague was not even the most common cause of death in the fourteenth century. Most deaths, especially in children, were from other forms of infectious disease. Plague can most plausibly be seen as a result of generally poor resistance to disease.

During this period, death was unusually common at all levels of society. In the fifteenth century, 10% of undergraduates died during their period at university. Infant mortality was also high, with many noble houses dying out because they were unable to rear sons. Neither of these groups were ill fed, and noble families tended to live on country estates which were much healthier than the towns.[170]

Nor can the high death rate among ordinary people be explained by poverty. Wages were considerably *higher* in the low-growth fifteenth century than in the periods before or after.[171] It is hard to avoid the conclusion that people were simply more susceptible to disease, and this is entirely consistent with a G-150 year of 1410.

But the decline in population was not only due to death. People were having fewer children, as were muskrats in their decline phase. Besides the failure to produce surviving heirs, some noble families in the fifteenth century simply did not produce children. This appears to have been in part voluntary, as if during the decline phase the desire to reproduce simply wasn't there. For example, *coitus interruptus* was commonly used in the fourteenth century.[172]

We see a similar overall pattern in the demographic slump of the late seventeenth and early eighteenth centuries. It was a relatively prosperous time, and yet the population of England grew very little. Not only was the birth rate lower than in the centuries before and after, but the death rate was also higher (see Fig. 8.5 below).

These facts defy conventional socioeconomic explanations. Why should people have fewer children in more prosperous times? Even the wealthy, who were fully able to support more children, gave birth to fewer of them. Although difficult to explain by other means, this evidence is exactly what we would expect from the lemming cycle pattern in its decline phase, with G-150 periods around 1410 and 1700.

Changes in behavior and temperament also correlate with lemming cycle phases. Periods around the G year have the energy and confidence of being close to the peak of V, but they also exhibit something else not necessarily predicted by the theory—a strong and vibrant sense of national unity based on patriotism rather than fear. In other words, people in G periods become extremely positive about their nations and want to support their leaders.

Fig. 8.5. Crude birth rates and death rates (per 1000) in England, 1541–1871.[173]
As with muskrats, human birth rates are highest and death rates lowest in the G (growth) stage of the lemming cycle. The opposite applies in the low G stage, at G-150.

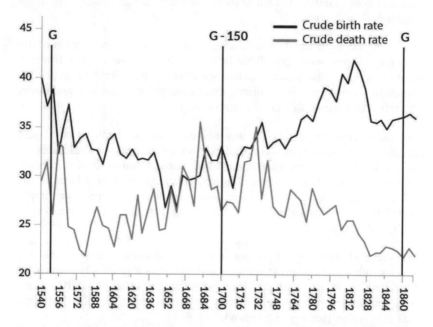

Going back beyond the demographic data, we can identify an earlier English G year around 880 AD, when Alfred the Great rallied the Saxons to defeat the Danes and form a powerful, centralized kingdom. By 940 AD, the G+60 year when local loyalties are expected to be weakest and acceptance of remote authority strongest, Alfred's grandson Athelstan completed the unification of England by capturing York in 927.

From the G+60 year of 940, England grew steadily weaker, allowing first Danes and then Normans to take control. The first Norman kings were powerful, as might be expected when the entire ruling class was replaced, but between 1135 and 1150 the country was plunged into civil war as turbulent barons switched sides and "God and his angels slept." This period coincides exactly with the G-90 year of 1140, the peak of infant C and low point of child V. Based purely on the demographic pattern, this is the year we would *expect* local loyalties to be strongest and autocracy weakest. It is easy for barons to rebel when they know their soldiers will follow them regardless of who claims to be king.

From then on national unity was rapidly restored, consistent with another G year in 1230. Powerful monarchs such as Henry II, Richard I and Edward I were once again in control. Rebellions were quickly put down and the economy grew. The peak of central control was during the reign of Edward I (1272–1307), a fearsome autocrat who subdued both Wales and Scotland. His reign brackets the G+60 year of 1290, when child V is high and infant C low so we would expect autocracy to be strongest and local loyalties weakest.

This had two main effects. Weak local loyalties gave Edward's nobility little scope to resist, allowing him to collect heavier taxes than previous kings. It also meant weaker national loyalties, with the result that many Welsh and Scottish nobles freely supported him. Robert the Bruce, who set Scotland free after Edward's death, had been a supporter of England during most of his reign. And when Edward invaded Wales in 1277 more than half his force was Welsh.

Based purely on demography we would expect another G-90 period around 1460, characterized by strong local loyalties and weak autocracy. And again, this is exactly what happened. From its thirteenth century peak the monarchy gradually weakened, reaching a low point in the late 1450s during the Wars of the Roses. Nominally, these conflicts were over the crown, but beneath them was a maze of local disputes. Ambitious men with obedient retainers took the opportunity to seize neighboring lands, gaining support by aligning themselves with one or another royal faction. As discussed in chapter one, a typical example of such behavior can be found in Richard Neville, the Earl of Warwick. He switched allegiance from the house of Lancaster to the Yorkists, then switched again and restored the Lancastrians (briefly) to power, earning his title of "Kingmaker." In a similar way, Lord Stanley decided the battle of Bosworth Field in 1485 by throwing his support at the last minute to Henry Tudor, whose claim to the throne was remarkably weak. This change of sides occurred during the actual battle, and it was never doubted that Stanley's followers would fight for whichever claimant he chose.

During the next cycle, with its G year in 1550, another line of powerful monarchs ruled with little challenge to their authority. Henry VIII (1491–1547) was the most powerful monarch in English history and his daughter Elizabeth (1558–1603) perhaps the greatest. In this cycle the lemming pattern harmonizes with the civilization cycle, the G period coinciding with the latter's peak of V and stress and greatly enhancing the power and security of the Tudor rulers. An even clearer sign of weak local loyalties is the union of the English and Scottish thrones in 1603, very close to the G+60 point of 1610 when child V was strongest and local loyalties weakest.

By the next G-90 year in 1760, executive power had weakened again. However, the long-term upward trend in the level of C meant that there was no return to feudal chaos. Instead, local and hereditary loyalties were reflected in the overwhelming power of rural landowners, especially aristocrats. The same web of patronage and loyalty existed as in the Wars of the Roses, but exercised more peacefully due to higher levels of C. Among many other measures that favored the aristocracy, a ban on imported wheat kept food prices and thus farm profits high. Their position was maintained by a parliamentary gerrymander that put many seats under the control of local landowners (so-called 'pocket boroughs' and 'rotten boroughs') while the growing towns were under-represented. But as England moved towards the G year of 1850, aristocratic power declined. This was reflected in the Great Reform Act of 1832, which made electoral districts fairer, and the repeal of the Corn Laws in 1846.

By 1850 Britain had become a world power, with a high degree of vigor, optimism and national unity characteristic of G periods. By this time, the English civilization cycle had reached a point where "local" loyalties might as well be thought of as equivalent to national loyalties. Nationalism was then undermined by the weakening local loyalties of the next G+60 period in 1910. Just nine years later came the founding of the League of Nations.

Overall, then, there is a remarkable correspondence between the G periods of the lemming cycle—which represents demographic trends—and shifting political loyalties in English history, all working within the longer term trends of the civilization cycle. But there are various other observations which do not fit quite as well with the initial model.

From this we might expect economic growth to be strongest in the G-120 year at the peak of C, but in fact economies tends to do best in G periods.

That may be partly because of a more ordered government, but more likely has to do with the general temper of energy and confidence. It is also likely that the absolute change of C in the lemming cycle is quite small, which is why it does not produce noticeable changes in economic performance. The higher level of C at G-120 may, however, explain why the age of marriage in England was later in the early eighteenth century rather than the nineteenth at the peak of C, a problem noted in the last chapter.

Creativity and innovation also follow a slightly different pattern from what we might expect. Openness to new ideas should peak at the G-90 period when infant C is high and child V low. But in fact it peaks in the period *between* G-90 and the G year, when (perhaps) infant C innovation combines with the dynamism of the G year. Such times are marked by new ideas and rediscovery of the past. "Renaissance" has been applied to the twelfth century following the G-90 year of 1140, and the early sixteenth century following the G-90 year of 1460. The term "Enlightenment" has also been applied to the late eighteenth century, following the G-90 year of 1760. The opposite G+90 period of the cycle is associated with religious orthodoxy.[174]

The other aspect of the lemming cycle not yet considered is the aggression and migration that might be expected of the G+30 period when V is at its peak, a topic to be considered in the next chapter.

The most striking feature of the lemming cycle in English history, and which will be seen as a feature of all human history, is the relatively slow decline into disorder at the G-90 period, followed by a rapid resurgence of central power as the G year approaches, and then a peak of absolutism and low point of local loyalties sixty years later. This pattern will allow us to track lemming cycles in other societies, even in the absence of demographic evidence.

Lemming cycles in Japan

It is also possible to plot lemming cycles in Japan over more than a thousand years (see Table 8.2 below). Population seems to have grown substantially in the late sixteenth and early seventeenth centuries, slowing in the eighteenth and then rising in the late nineteenth and early twentieth centuries to a peak in 1925. This suggests lemming cycle G periods of 1590 and 1925. Putting this together with political data, Japanese lemming cycles can be traced back to the seventh century BC.

Table 8.2. Lemming cycles in Japan[175]

G-90	Period leading up to the G year	G year	Time since last G year
630	Taika Reforms— adoption of Chinese culture	720	
1110		1200	480
1500	Christianity, open to West	1590	390
1835	Opening to the West	1925	335

As in England, the G years display national unity and vigor. There is also an openness to change immediately before the G year which gives way to reaction and tradition after the peak of child V is reached at G+60. This accounts for changing Japanese attitudes to the West, as can be seen in the upheavals of the sixteenth and early seventeenth centuries.

The sixteenth century started with the Onin War, a time of feudal anarchy bracketing the G-90 year of 1500. From then on Japan moved rapidly towards unification, which was achieved in the G year of 1590. The years preceding the G year were a time of opening to the West and widespread adoption of Christianity. But as the new Tokugawa regime consolidated power in the early seventeenth century and approached the G+60 year of 1650, there was a severe reaction. Christianity was banned and contact with Westerners was limited to the single port of Nagasaki.

Approaching the next G year of 1925 there was another opening. The Japanese, who had been adamant in their rejection of all foreign influences for two and half centuries, became enamoured of Western technology, science, political institutions and even clothing. This opening-up can be traced to the year 1868 when rebel forces overthrew the weak Tokugawa regime and established a new and more powerful government. In other words, just 57 years before the G year.

This G year was *not* followed by a conservative reaction, because it was succeeded by a rapid fall in C related to the civilization cycle, but in all other ways Japan fits the lemming cycle pattern precisely.

China

China is rich in detailed historical information going back three thousand years. It is also helpful that Chinese history focuses strongly on dynasties, which is how most Chinese lemming cycles express themselves.

Traditionally, Chinese historians have identified a repeated pattern of dynastic change, in which a powerful new dynasty gradually decays into chaos, followed by a rapid resurgence of central authority. They saw it as the key theme of history, with the decay of central power reflecting the moral decline of successive emperors until the "Mandate of Heaven" was lost. Biohistory also sees dynastic cycles as reflecting changes in character, but in the people as a whole rather than the ruling house. This same pattern is identified in England and Japan, with the G-90 period representing the chaos, and a powerful new dynasty arising as the G year approached. Chinese lemming cycles are identified in Table 8.3 below.

Table 8.3. Lemming Cycles in China[176]

G-90 year	G year	Length in years
560 B.C.	470 B.C.	
300 B.C.	210 B.C.	260
10 B.C.	80 A.D.	310
560	650	570
930	1020	370
1330	1420	400
1630	1720	300
1920	2010	290

The history of China in the twentieth century shows a classic lemming cycle pattern. Following the previous G year of 1720, when the new Qing dynasty had been at the peak of its power and population grew rapidly, central authority slowly declined. By the mid-nineteenth century China was weak, humiliated by unequal treaties and a war in which British gunboats forced the Chinese government to allow the import of opium. At the same time, the Taiping rebellion—a massive political and religious civil war in the Yangtze valley—cost tens of millions of lives.

But the absolute nadir of central power in China came when the rule of the Qing dynasty dissolved into the Warlord Era of the 1920s, which is a perfect example of a G-90 period around 1920 when local loyalties should be strongest. China collapsed into complete chaos, with shifting coalitions of warlords fighting for power and no vestige of central government. A study of Chinese warlord armies in the 1920s shows how the political instability stemmed from psychological attitudes.[177] There has been no hereditary landed nobility in China for well over a thousand years, but warlord armies showed the same intensely local and personal loyalties as the people of fifteenth-century England. Soldiers were loyal to their own commanders and cases of individual treachery and wilful desertion were rare, except when an entire unit changed sides under direct orders of its commander. This is exactly what happened during the Wars of the Roses when the Earl of Warwick and other notables changed sides.

It was this that made the politics of the time so confused. Common soldiers owed allegiance to their own local leader, not to the warlord himself. The local commander had a personal tie with his own superior, the warlord, who might in turn support one of the most prominent national figures. Because of this, a defeated warlord might be quickly abandoned by his followers.

The power of the warlords also depended at least in part on personal support from the broader community. A 1923 poll of students and businessmen—the least likely groups to be supportive of warlords, one would think—found that no fewer than five of the most admired twelve living Chinese were warlords. Even people who deplored the chaos of the country often spoke of individual warlords with respect.

From then on, China went through the classic lemming cycle pattern of a swift rise to national unity and power. The Nationalist government of the 1930s regained some authority, though they were never able to fully subdue the entire county. And then, in 1949, a new and powerful Communist government took control, regaining surrounding areas such as Tibet as their Qing predecessors had done centuries earlier. Once again this is a typical lemming cycle pattern, with a swift rise in central authority following the slow decline.

China in the early twenty-first century has all the vigor, confidence and unity of a typical G period around 2010, even though population growth is relatively low as a result of the one-child policy. This is a vivid confirmation of the principle observed in muskrat populations—that cyclical changes in attitude persist even when the actual sizes of populations are quite different. For example, local muskrat populations "peak" and "collapse" in line with the ten-year cycle, even when local populations may be quite low as a result of (for example) a drought.

The radical reshaping of Chinese society by the Communists is also typical of the openness to new ideas that is a characteristic of the half-century or so before the G period. Such openness has continued into the twenty-first century and shows no sign of fading as in a normal lemming cycle, but (as in Japan) this is because of collapsing C, a topic we will come back to in later chapters.

Why do lemming cycles vary in length?

A working figure of three hundred years has been used as the expected length of a human lemming cycle, and for English history this works reasonably well, but not perfectly. The English G years are at 880, 1230, 1550 and 1850, which means that they progressively shorten—350 years, 320, 300. Japanese cycles show the same but rather more pronounced pattern—480, 390, 335.

In Chinese history the differences in length are the most striking of all, from a high of 570 years to a low of 260, but the overall pattern is the same. Cycles are shorter in times of rising civilization, such as from 560 BC to 80 AD and from the seventeenth century up to the present. The cycles are longer in times of disorder, such as in the Chinese "Dark Age" between 200 and 600 BC. The transition in length is smooth, except for the 1020–1420 cycle when China was overrun by Mongols and other barbarians.

A final important point is that this stretching and contracting of the cycle does not happen evenly to all sections. The interval between G-90 and G year seems always to be ninety years, as does the interval between the G year and G+60 year. In other words, the extended section of the cycle is the one in which child V is falling. This implies that the fall of V in the lemming cycle takes longer when stress in the civilization cycle is rising.

The biological function of lemming cycles

If the potential for lemming cycles is universal in mammalian populations, as these human examples indicate, it may be worth considering their biological function. We have seen that adapting to a variable environment requires a great deal of migration, so that if populations die out in one area they can survive and then flourish somewhere else. This is why high V populations are unusually sensitive to crowding stress, so that they migrate readily even when food is still abundant locally.

Lemming cycles are another kind of adaptation to unstable environments. To escape from predators, small animals must be timid and cautious, sticking to familiar areas where they can easily find refuge. Baboons can gang up on and occasionally kill leopards, but even the most ferocious rabbits could not take on a wolf or eagle. Yet timid animals attached to their home territory will not be good at migration.

A cycling population solves this conundrum. Normally timid and shy, especially when numbers are low, they can readily hide from predators. But every few years their numbers explode, much faster than slower-breeding predators can match. This temporary advantage makes it possible to breed a population cohort which is both large and bold. Because they are so numerous relative to predators, their boldness does not cause undue mortality. Boldness and sensitivity to crowding means they scatter widely and take up all available ecological niches, after which numbers collapse so that predators starve. If it were not for the collapse, the predators would remain numerous and the next bold generation would be comprehensively eaten. This theory explains why population cycles are greater and more frequent in harsher northern regions.

Humans simply share this mechanism with our mammalian cousins, even though it has less value in our case given that we breed more slowly than our major predators, such as lions.

Conclusions

Essentially, the lemming cycle is an alternation between V and C, driven by the rise and fall of a still-mysterious variable labelled "G." This variable is strongly linked to population growth, to the extent that it dwarfs the effect of C in the lemming cycle and causes a significant perturbation in population, political, economic and other trends in the civilization cycle.

Testing

A population of mice or rats in a naturalistic environment, with plentiful space and variable food supplies, should show lemming cycles around ten generations in length. These would be identified by changes in litter size, age of breeding, boldness, disease resistance, etc. Detailed physiological studies would help to make clear the characteristics of different stages of the cycle. Cycling animals in the wild should have epigenetic signatures appropriate to each stage of the cycle.

CHAPTER NINE

WAR

I can calculate the motion of heavenly bodies, but not the madness of people.
—Isaac Newton

In chapter four we saw that warlike human societies are those with a higher level of V, which is largely a result of highly anxious mothers transmitting this anxiety to their infant sons. The previous chapter explained lemming cycles as a ten-generation alternation between V and C, suggesting that societies might be more warlike at the peak of V. This chapter will apply these notions to the subject of war, explaining not only *why* wars break out but (to some extent) *when*.

In all of human history, there can be no subject as compelling as war. Historians have spent more ink on charting it than on any other aspect of the human past. Individual wars generate hundreds, if not thousands, of new books every year.

One of the perennial questions about wars is how they start. Not just the precise circumstances and causes of individual wars, but the origins of war itself. Attempts have even been made to trace a universal pattern of causal factors underlying all wars. Such attempts have not generally been very successful, probably because the political, military, demographic, economic and sociological factors that historians have to deal with are so intensely complex as to defeat such analysis.

Not so with biohistory. As we have seen, by founding our understanding of civilization and history on biology, it is possible to identify underlying patterns which seem to be, for all practical purposes, universal.

The basic idea of biohistory is that war, like other developments in the course of a civilization's history, is driven by temperament. And the system most relevant to war is V, which is characterized by vigor and aggression as well as the urge to migrate, expand one's territory, and dominate one's neighbors.

The Great War of 1914–18 was a conflict without equal or precedent in modern European history, rivalled only by the Napoleonic Wars of more than a century earlier. One of the Great War's defining features is the mass enthusiasm that surrounded its outbreak. In many of the belligerent nations—particularly Great Britain and Germany—the declaration of war in August 1914 was greeted by a display of jubilation that encompassed all social classes. In London, Munich and Berlin crowds poured onto the streets to cheer their leaders. The war was an exciting adventure, and young men rushed to enlist for fear that it would be over too soon and they would miss out. This is how one young German viewed the outbreak of war:

> A gigantic wave of fiery hot feeling passed through our country flaming up like a beautiful sacrificial pyre. It was no longer a duty to offer one's self and one's life—it was supreme bliss. That might easily sound like a hollow phrase. But there is a proof, which is more genuine than words, than songs, and cheers. That is the expression in the faces of the people, their uncontrolled spontaneous movements. I saw the eyes light up of an old woman who had sent four sons into battle and exclaimed: "It is glorious to be allowed to give the Fatherland so much!"[178]

In Britain, men volunteered in their millions, at one stroke adding hundreds of new battalions to the army. In Europe, millions more responded to conscription with little protest or resistance. They became the soldiers who surged out of the trenches in vast human waves, tangled in barbed wire and shot down by enemy machine guns. Casualties, especially among junior officers, were at levels no Western army would tolerate today. Peace came in 1918, but lasted only twenty-one years before international conflict broke out again. This time the results were even more catastrophic. During the global war that followed more than sixty million people died from combat, bombing, disease, starvation and genocide.

How can we explain these two conflagrations—the first entered into with such enthusiasm, leading to destruction and trauma which somehow failed to prevent the second from happening? The answer lies in testosterone, in V, in the lemming cycle, and in stress—all topics covered in previous chapters.

In considering the origins of war there are four causes at work. The first is that wars are more likely when a rapid drop of C in the general population produces a surge of testosterone. The second, as indicated in chapter four, is that wars are more likely in patriarchal societies that make women anxious and give men higher status, which maximizes V. The third is that wars of aggression are more likely to occur during the high-V phase of the lemming cycle (G+30), as discussed in the previous chapter. The fourth is that wars are more likely when mothers are made anxious by a previous war, and transmit this anxiety to their infant sons. Male offspring born at that time, but growing up in peace, will have especially heightened V.

We will consider testosterone first.

Testosterone surges from falling C

Experience of C-promoters (such as food restriction) in infancy does not reduce testosterone and may even increase it, but the same C-promoters in later childhood and adult life actually *reduce* testosterone (see chapter three). This means that people who reach maturity at a time when C is falling should have higher testosterone than other generations. Furthermore, the level of testosterone should be affected by the speed with which C is dropping.

C has been falling in Western societies for the past 150 years since its peak around 1850. There are many strands of evidence for this, one of which is the declining age of puberty. Age at puberty is a good indicator of C—the later the age, the higher the level of C. In late nineteenth-century Europe it was unusually late, but by the end of the century it had begun to drop quite rapidly (see Fig. 9.1 below).

In Germany the age of puberty reached an all-time peak in the 1870s and then fell rapidly. Britain experienced a substantial though lesser decline, and the French decline, apart from an anomalous result in the 1890s, was more gradual still. The decline was also relatively slower in Sweden. What is interesting is that neither France nor Sweden displayed the same upsurge in martial spirit in the early twentieth century.

On this evidence, the surge of testosterone in early twentieth-century Europe should be greatest in Germany. Judging by the still more rapid fall in the German age of puberty in the 1930s, the generation maturing in that decade should also have had unusually high levels of testosterone.

In terms of testosterone levels, then, the European nations with the largest, most rapid increase were those most enthusiastic about the declaration of war in 1914. And the instigator of war in 1939, Germany, had the biggest rise of all.

Fig. 9.1. Age of menarche in Germany, Great Britain, and France.[179] A rapid fall in C, as indicated by a rapid decline in the age of menarche, brings C below infant C. This creates a surge of testosterone and thus aggression, especially in Germany where the age of menarche fell faster than in most countries.

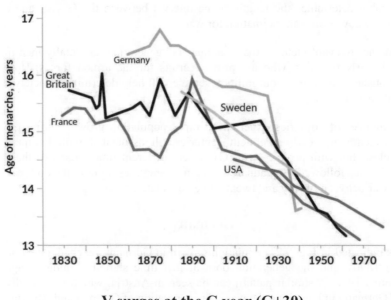

V surges at the G year (G+30)

The second cause of war is patriarchy, which has been dealt with at length in previous chapters. The third cause is lemming cycles which are, in essence, multi-generation fluctuations between C and V. In one half of the cycle C is rising, in the other it is V. These fluctuations occur within the larger, longer-term rises and falls of the civilization cycle.

The V peak of the lemming cycle occurs thirty years after the peak of population growth or "G" point ("G+30" in lemming cycle notation). In lemming populations this is the time when these normally timid and unadventurous animals become bold and set out on the mass migrations for which they are known. In human societies, a surge of aggression may be expected when the generation born around the peak of population growth in the G year comes to maturity. This generation would have unusually high V.

How much of an effect this would have on a society's warlike character depends on how influential that generation is, and at what age it becomes influential. In some societies the most influential generation might be in their thirties or older, but in the recent West there is evidence that people in their early twenties have a powerful influence on the attitudes of society as a whole (see the next chapter's analysis of the onset of recessions). This variable determines the length of the interval between the G year and a society developing an inclination for war.

Another relevant factor is that it is this "G" generation, especially men in their early twenties, who are prime recruits for the armed forces. How enthusiastic they are about military service will help determine a society's military capability.

A number of countries experienced rapid population growth during the nineteenth and early twentieth centuries, all associated with lemming cycles, but with peak growth (G) at very different times. Each of these peaks was followed by gradually rising militancy, resulting in the outbreak of war between twenty and twenty-five years later.

Germany

As indicated in Fig. 9.2, population growth in Germany reached a peak in the 1890s, and the generation born at that time started coming of age around 1910. A similar pattern can be seen in Austria, which was an ally of Germany in launching the war. This generation of German and Austrian men would also have had very high levels of testosterone.

Ernst Junger, who was to be seriously wounded seven times and fought throughout the entire war, expressed the incredible fighting spirit of the Germans:

> We had grown up in a material age, and in each one of us there was the yearning for great experience, such as we had never known. The war had entered into us like wine. We had set out in a rain of flowers to seek the death of heroes. The war was our dream of greatness, power, and glory. It was a man's work, a duel on fields whose flowers would be stained with blood. There is no lovelier death in the world ... anything rather than stay at home, anything to make one with the rest.[180]

Even four years of fighting, for many on the German side, could not wholly extinguish the lust for battle. This was how Junger described the March 1918 German offensive:

Fig. 9.2. Rate of natural increase per 1,000 (5 year rolling average) in Germany. The high V generation born at the peak of population growth (G) came of age in 1914, reinforcing a mood of aggressive nationalism that was also driven by a testosterone surge as a result of falling C.

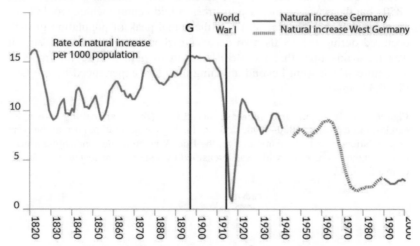

> The turmoil of our feeling was called forth by rage, alcohol, and the thirst for blood as we stepped out, heavily and yet irresistibly, for the enemy's lines. And therewith beat the pulse of heroism—the godlike and the bestial inextricably mingled. I was far in front of the company, followed by my batman and a man of one year's service called Haake. In my right hand I gripped my revolver, in my left a bamboo riding-cane. I was boiling with a fury now utterly inconceivable to me. The overpowering desire to kill winged my feet. Rage squeezed bitter tears from my eyes.[181]

The most striking aspect of this war was the apparent lack of rational calculation. Germany had launched a short, sharp war with France in 1871, a campaign which was swift and victorious. France was humiliated and two border provinces seized. Germany gained immensely in terms of territory, power and prestige with minimal casualties and risk. By contrast in 1914 (as again later in 1939) Germany took on nations with vastly greater resources of manpower and wealth. It is as if the warlike temperament of the time blew away any sensible calculation of national interest. The results were catastrophic, with two generations of young men slaughtered, civilians dead, cities ruined, national humiliation, occupation and territorial dismemberment.

Britain

Britain also experienced a surge of population growth in the nineteenth century. The G year, according to calculations, would have been around 1850, but there was an increase in stress at mid-century which would have suppressed V (see chapter ten). So, the actual peak of population growth occurred during the 1870s, two decades earlier than in Germany. And it was from this time that the British began to gain a taste for imperial adventure which went beyond anything they had experienced before (see Fig. 9.3 below).

Fig. 9.3. Rate of natural increase per 1,000 (five-year rolling average); England and Wales: 1800–2000.[182] When the high V generation born at the peak of population growth (G) came of age, the Boer War broke out, indicating a peak level of aggression. Aggression was intensified by a surge of testosterone resulting from a fall in C.

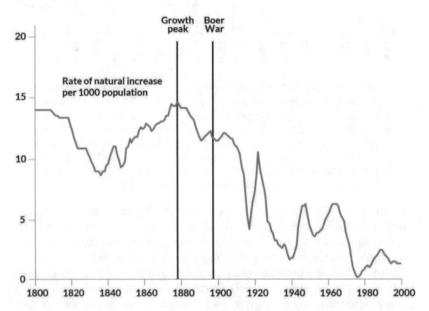

The British imperialism of the late nineteenth century was of a markedly different order from anything that had happened earlier. Earlier in the century, British interest in its colonies had been mainly commercial and practical. Britain's Far Eastern possessions were owned by the East India Company, which provided its own armies and was regulated by the British parliament. The mood in Britain was one of wishing to limit the acquisition of new overseas territories due to the administrative and moral headaches they tended to cause. But in the aftermath of the 1857-58 Indian Mutiny, the Government of India Act dissolved the Company and gave the UK government full and direct control over all its possessions. At around the same time, the mood in Britain began to alter.

From the 1870s onward, British politicians such as Benjamin Disraeli gained popularity by appealing to the sentiments of empire. The aggressive ambition of men like Cecil Rhodes gave a vast boost to the British Empire's dominance in southern Africa, an expansion which drained national income and made very little commercial sense. In his autobiography of the decades before the Great War, Esmé Wingfield-Stratford commented on the rampant militarism among the young, highlighting the mood of the time:

> [T]he whole atmosphere of the time seemed to be faintly redolent of gunpowder; ... among those who professed and called themselves gentlefolk in the fin de siècle—and I think this would apply to an even wider circle—everybody seemed to be talking about those two linked attractions of war and empire.[183]

The strongest expression of this new mood was the Boer Wars of 1899–1902, which caused immense suffering and loss of life, and not only among the soldiers. Tens of thousands of Boer women and children died from hunger and disease in British concentration camps. South Africa gained its independence a few years later, so apart from the retention of gold-mining interests there was little practical benefit to the British from the struggle. It could be regarded as the most irrational war in modern British history and it came just twenty-five years after the peak of population growth, when a generation with high testosterone and maximum V came of age. By 1914 the fervour for war had ebbed slightly but was still at a very high level. This was only fourteen years after the absolute peak of V, and combined with the early twentieth century testosterone surge to make Britain willing and ready to take on the Germans.

Japan

In the decades following the Meiji Restoration Japan built itself from feudal backwater to industrial powerhouse. After crushing the Russian navy in 1905 it seized Korea and Taiwan, becoming the dominant military power in the region. All this came at immense cost to the conquered peoples but with little risk or danger to the Japanese. Like the Franco-Prussian war, it was relatively rational in terms of the achievability of its objectives.

Then, after about 1910, the Japanese age of puberty began to fall. The decline was much less than in Germany or Great Britain, but Japan was a far more patriarchal (and thus higher V) society to begin with, so the underlying potential for aggression was greater. At the same time, Japan experienced a surge of population growth, reaching an all-time peak in the 1920s (see Fig. 9.4 below).

Fig. 9.4. Rate of natural increase per 1,000 (five-year rolling average), Japan 1870–2000.[184] Japan reached a peak of population growth (G) in the 1920s. The high V generation born at that time fought in the Pacific War. The testosterone surge from falling C reinforced their aggression.

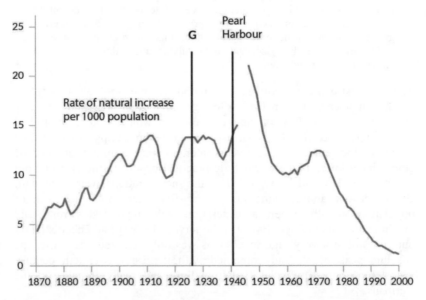

From this time onwards, the Japanese taste for imperial conquest really took hold. In 1931 they occupied Manchuria. In 1937 they launched an all-out assault on the rest of China, attacked Pearl Harbor in 1941, and began a swift occupation of Southeast Asia. It was a hugely ambitious project and, as with Germany in 1914 and 1939, the fervour for war was powerful enough to override all practical objections. In fact, many of those who spoke out against the war were assassinated by nationalist fanatics. The

Japanese took on nations with larger populations and far greater industrial production. The result for Japan, as in Germany, was the death of millions of people, mass destruction, occupation, and the loss of empire.

The same pattern can be seen in other countries.[185] France reached a peak of population growth at the end of the eighteenth century. It was the first European nation in this era to reach the G year on the lemming cycle, and the first to launch a campaign of conquest, leading to the Napoleonic Wars. As with Germany and Japan in the twentieth century, France took on nations with vastly greater populations and resources and at a huge cost to itself. Similarly, Italy reached its own growth peak in 1915 and invaded Ethiopia twenty years later in a quest for imperial glory. It then joined Germany in the suicidal folly of the Second World War.

United States

Prior to the twentieth century, the United States engaged in several major wars. The two most significant were the War of Independence of 1775–83 and the Civil War of 1861–5. Because of its unique geographic and demographic situation, the pattern of lemming cycles in North America is different to that of Europe.

Precise population figures are not available for the eighteenth and most of the nineteenth centuries, but we can note some trends. It is known that British North American colonies experienced rapid population growth in the late colonial period, increasing by as much as 700% between 1689 and 1760.[186] Much of this was a result of immigration, but a large part was from natural increase. The colonists were quite well off compared to people in Europe—land was freely available and food was cheap, and there were none of the famines common in Europe.[187] The *average* annual rate of natural population increase between 1720 and 1800 seems to have been around 25 per 1,000. European countries only achieved this briefly, at lemming cycle peaks.[188]

Such a fast-growing population suggests a very high level of V, supported by the extraordinary willingness of ordinary people to take up arms during the rebellion against the British. In September 1774, upon hearing a rumor that Boston had been bombarded, and fearing the seizure of powder stores, a vast gathering of New Englanders (estimates range from twenty to sixty thousand) left their homes and headed towards the action.[189] According to one eyewitness:

> For about fifty miles each way round there was an almost universal
> Ferment, Rising, seizing of Arms & actual March into Cambridge ...
> [T]hey scarcely left half a dozen Men in Town, unless Old and Decrepit.[190]

Supporting the war effort, Connecticut's committee of correspondence
(the provisional state governments set up to coordinate the colonies'
resistance to the British) declared: "The ardour of our people is as such,
that they can't be kept back." Within a week, 3,716 men from Connecticut
were marching to the aid of Massachusetts, and by early summer an
estimated twenty thousand revolutionaries had surrounded Boston,
trapping the British soldiers in the confines of the city.[191]

A more typical example of a G+30 war, however, can be found in the
Civil War.

The American rate of natural population increase remained strong through
the first two decades of the nineteenth century, perhaps even rising
slightly, but from the 1820s it began a steady fall. This suggests a G period
for the United States as a whole around 1825 (see Fig. 9.5 below).

It is characteristic of lemming cycles, as demonstrated by the European
nations of the nineteenth and twentieth centuries, that expansionist
sentiment rises rapidly from the G period to a peak around twenty-five
years later. The same thing happened in America. Expansion into new
lands was driven by the sense of "Manifest Destiny," as popularized by
John O'Sullivan in 1845:

> ... that claim is the right of our manifest destiny to overspread and to
> possess the whole continent which providence has given us for the
> development of the great experiment of liberty and federated self-
> government entrusted to us.[192]

Fig. 9.5. Rate of Natural Increase per thousand, United States 1790–1990[193]

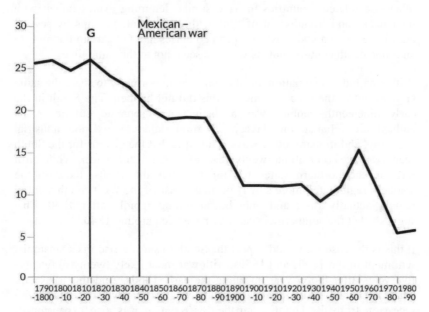

Even those who were opposed to the policy of expansion held a sort of fatalistic resignation that some higher power was directing the nation. In 1837 the Boston preacher William Channing wrote an open letter to Henry Clay, criticizing the motivation for expansion:

> We are destined to overspread North America; and, intoxicated with the idea, it matters little to us how we accomplish our fate. The spread, to supplant others, to cover a boundless space, this seems our ambition, no matter what influence we spread with us. Why cannot we rise to noble conceptions of our destiny?[194]

Many saw the expansionist zeal of this period as part of a universal plan in which Providence was working behind the scenes to lead the nation toward its intended destiny.[195]

In the northern states, this sentiment reached a peak almost exactly 25 years after the peak of G—the Mexican-American war of 1846–8, when America seized lands from Mexico equivalent in size to the whole of Western Europe.[196]

But this is not the entire story. There is a tendency for neighboring (or otherwise related) countries to have similar lemming cycles, albeit with their peaks and troughs out of sync with each other, perhaps by several decades. America and most European countries had G years in the mid- to late-nineteenth century, but, as we have seen, not at the same time.

Within an individual nation, local regions seem normally to share the same G year, but in the case of America this did not happen. The South in the early nineteenth century was a distinct and separate culture in its enthusiastic reliance on slavery, its rural demography, non-industrial economy, and in many other ways. Thus it is that the G year for the South seems to have come about twenty years after that of the North. While birth rates in the northern states fell rapidly after the 1820s, those in the southern regions fell more slowly or, in the case of the West-South-Central region, actually rose, and only began falling rapidly after 1840. This suggests that the southern G year may have been around 1840.[197]

If this is the case we would expect the South to show a rise in expansionist sentiment in the 1840s and 1850s, with war most likely twenty to twenty-five years after the G year, meaning the early 1860s. And this is exactly what happened. Although the South was primarily concerned with secession from the Union from the 1850s on, it was also expansionist. There was a powerful sentiment for spreading slavery and the South's version of Americanism into the western territories and the tropics, now reclassified as a paradise. One Southern congressman commented on expansion:

> With swelling hearts and suppressed impatience they await our coming, and with joyous shouts of "Welcome! Welcome!" they will receive us.[198]

And so it was that twenty-one years after the Southern G year, at an absolute peak of V in what was already a very high-V society, the South launched its war for independence. Like the Napoleonic Wars and the First World War, it was characterized by mass citizen armies and extraordinary fighting spirit. The first of the great wars of the modern era, the American Civil War carnage was made even worse by the new industrial technology created as a result of high infant C. It was the deadliest war in American history, costing as many casualties as all other wars combined, including 6% of Northern males and an incredible 18% of Southern white males between 13 and 43.[199]

The fighting spirit of the Confederate soldiers was extraordinary. Though far outnumbered and desperately short of food, ammunition and even clothing, they kept the struggle alive for four years, at times even invading the North.[200] In battles such as First Manassas and Second Bull Run, outnumbered Southern armies achieved notable victories. This kind of martial spirit is characteristic of nations in the G+30 period, including the armies of Revolutionary France and of Germany in the Great War.

Decisions about war and peace are made at all levels of society and in all age groups, especially in a democracy. Political and military leaders on the whole tend to be much older men. And yet, historical evidence suggests that wars often break out when a generation born at a peak of population growth reaches peak military age, normally their early twenties. Surprising as it may seem, the warlike fervour of this generation somehow affects the entire society, or at least provides an unusual opportunity for aggressive and warlike leaders.

Finally it may be noted that most countries experience a surge in birth rate and population during the period after a major war, probably because of births delayed during the conflict. In Japan, for example, birth rate and population increase were a lot higher after 1945 than at any point during the 1930s and 40s. There were also post-war baby booms in the 1950s and 1960s, which are attributable to higher levels of child and adult C arising from the privations of the Great Depression. Yet neither of these population surges resulted in increased aggression once the age cohort matured. For example, American baby-boomers born in the 1950s and early 1960s came of age in the 1970s and early 1980s, when student activism was in decline and there was no taste for military adventure. Thus it seems that the V which drives these major wars is a product of the lemming cycle rather than of population growth as such.

Aggression arising from trauma experienced in infancy

We have said that there are four major causes for war. The first is a tradition of patriarchy, which causes women to be more anxious and to transmit this anxiety to their infant sons. The second is a surge in testosterone from rapidly falling C, as occurred in Europe in the early twentieth century. The third is a peak in V at the G+30 year of the lemming cycle. And the fourth is stress in early infancy as the result of an earlier conflict. It is this last we must deal with now.

As indicated in chapter four, the most aggressive individuals are those who experience a great deal of anxiety in infancy, transmitted by an anxious and indulgent mother, but who grow up to experience much lower anxiety in later life. This maximizes the level of V. The same applies at the level of society. When an age cohort is stressed in infancy but *not* later, that entire cohort would be expected to grow up with higher V—and thus be unusually aggressive.

This is exactly what happens to children born at the end of a major war but growing up in peace. A full-scale war is a hugely stressful event for most people. Families are subject to violent intimidation by occupying soldiers, or flee their homes in terror. People cower in cellars and shelters from bombardment by artillery or aircraft. In many cases, famine and privation follow as resources are diverted to the war effort or fighting blocks off supplies. On top of these stresses, mothers with infants are likely to have husbands involved in the conflict, and are in a state of persistent anxiety about them.

Then the war ends, usually quite suddenly when one side or the other surrenders. The fighting stops, people celebrate, the men return home and even for the defeated life usually begins to return to normal. The birth rate typically surges. In terms of promoting V, the greatest effect is on infants born in the last two years of the war who grow up in a time of peace. The more stressful the war and the more absolute the peace, the more aggressive should be the rising generation of young men. When they reach adulthood, if the theory is correct, we might well anticipate another war.

There is good historical evidence that wars commonly break out within twenty to twenty-five years of the end of a previous major conflict. In China, which has detailed historical records covering almost three thousand years, it is common for a war of unification to be followed by a second a generation later. China was unified by the Sui dynasty in 590 AD, but the new dynasty was destroyed by a massive rebellion which broke out in 614. The Ming dynasty came to power in 1371 and faced a massive rebellion in 1398. The Qing dynasty established control of China in 1651, and was almost destroyed by the "Revolt of the Three Feudatories" which broke out in 1673. These are gaps of between twenty-two and twenty-seven years. Other factors set the exact timing of the revolt, such as the death of a powerful leader, but it is uncanny how often this gap between wars is found.

This phenomenon provides an explanation for that greatest of tragedies, the Second World War. The Great War had been traumatic for all the parties involved, perhaps most of all for the defeated Germans and Austrians. There was a powerful sense of "never again," and the League of Nations was set up to help make sure of it. The anti-war culture that followed the war—in literature, films and politics—was powerful and widespread. And yet, only twenty-one years after the Armistice and with the memory of the slaughter still fresh, Germany once more embarked on a war of conquest, taking on nations and populations many times its size.

The rise of the Nazi Party is a different subject, discussed in chapter ten, but the lust for war can be seen as a direct product of the previous war. Boys born between about 1916 and 1918 would have felt the full effects of this anxiety in infancy, transmitted through their mothers, while experiencing lower anxiety in the post-war years. This super-aggressive generation reached their early twenties between 1936 and 1939, superseding the anti-war generation that preceded it, and providing Hitler's shock troops for the invasions of Poland and France.

The First World War was justified as the "war to end all wars," a claim which became one of the twentieth century's great ironic jokes. In fact, as we have seen, its effect was not merely that it failed to end war. To a great extent, it actually caused the next one..

Why no World War III?

The unprecedented aggressive militarism that scoured the world during the first half of the twentieth century resulted from a "perfect storm" of demographic and biological conditions. Family patterns that were still patriarchal combined with peaks of birth rate to produce a surge in V in the late nineteenth and early twentieth centuries. Later, the after-effects of the Great War produced another massive surge of V. Added to this was higher testosterone as C declined rapidly from the late nineteenth century in Europe, and in the early twentieth century in Japan.

The Second World War itself produced another generation of stressed infants growing up in peacetime. This higher V cohort both fought the Communists in Vietnam, and opposed the war vigorously with massive and often aggressive student protests. 1968 was also a year of violent student protests across Europe.

Similarly, the Chinese born at the end of the Civil War in 1949 were the teenagers who wreaked havoc on China in the Cultural Revolution unleashed by Chairman Mao in 1968. This experience in itself was stressful enough to create a further surge of student unrest culminating in Tiananmen Square in 1989.

This raises the question of why the world wars finished in 1945. If stressing infants raises V then the cohort born in 1944–45 should also be more aggressive and liable to launch another war. By this time, however, the other contributory causes were in sharp decline. As will be discussed in chapter sixteen, patriarchy and V in general declined rapidly in the late twentieth century. 1968 was nearly a century away from the last G year (depending on the nation). Also, the testosterone surge of the early twentieth century had very likely ebbed by this time.

It is comforting to realize, from this analysis, that major wars like those of the early twentieth century are *very* unlikely in the near future. The last major candidate was Iran, which experienced both a bitter war and a peak of population growth in the 1980s. A high V generation maturing around 2010 expressed itself in the relatively aggressive Presidency of Mahmoud Ahmadinejad in 2005-13, but Iran was never strong enough to be a real threat to its neighbors. Iranian militancy has since declined, as would be expected given that the current youth generation was born in peace time and when the birth rate was in decline. Unfortunately, there are other and equally serious dangers facing the West, which will be the subject of a later chapter.

Testing

It seems extraordinary that the temperament of a single generation of young men (and women) could have such a powerful influence that it launched their societies. But that is what biohistory proposes. And unlike any other theory of war, this one can be tested.

People born at the end of major wars (1944–45 in Europe or Japan, 1948–49 in China, 1987–88 in Iran) should have physiological indications of high V and be more aggressive than older or younger cohorts.

CHAPTER TEN

RECESSION AND TYRANNY

I trust no one, not even myself.
—Stalin

On September 3, 1929 the US stock market was at an all-time high. There had been an unprecedented bull run for nine years, with the Dow Jones Industrial Average worth ten times as much as it had been at the start of the decade. Some stocks did even better than that. In 1921, shares in RCA were worth $1.50; by 1928 they had soared to $420.[201] Stock speculation had become a national mania, with many investors buying "on margin" (using cash borrowed against collateral), which meant that a rise in the share price made big money but a fall could wipe them out.

Stock prices eased slightly during September 1929, but then on Black Tuesday, October 24, the market lost 11% of its value. Despite all the efforts of bankers and politicians the decline went on until the Dow was down 89% by July 1932. America was plunged into the strife and privation of the Great Depression. Industrial production fell by 46%, foreign trade by 60%, and unemployment rose to more than six times its previous level. Other countries caught the contagion, with unemployment more than doubling in France and Germany.[202]

Other areas of society, not just the economy, felt the effects. Political disputes turned bitter and sometimes violent. In an atmosphere of fear and paranoia, there was a growth of extremist parties on the right and left. Communists and Fascists battled each other in Spain. In Germany, the Nazis came to power. And while the origins of the Second World War can best be explained by the after-effects of the First (see the previous chapter), the genocidal viciousness of the Nazi regime can be directly traced to the poisonous politics of the era.

Recession cycles

Explanations for the Great Depression are many and varied. They range from Keynesian lack of stimulus to contraction of the money supply, to trade protectionism and attempts to freeze wages and prices. Biohistory, once again, takes a different approach. As with the cyclical patterns of civilization, the explanation of recession involves a change in the temperament of the general population. What is so striking about the period between the two world wars is the sharp contrast between the prosperity and optimism of the 1920s and the poverty and gloom of the 1930s. What biohistory suggests is that the prosperity actually *caused* the poverty and gloom.

Civilization cycles and lemming cycles both influence population growth. Civilization cycles cause it to peak at high C, and lemming cycles at high G. But there are shorter-term fluctuations that these cycles cannot explain, such as the slower growth of population in mid-nineteenth century Britain at what should be the peak of the lemming-cycle G period. America in the early twentieth century was in the declining phase of both civilization and lemming cycles and should thus show a steady decline in the birth rate. But as Fig. 10.1 shows, the decline was far from uniform.

Fig. 10.1. US crude birth rate 1910–2010 (births per 10,000 population)[203]

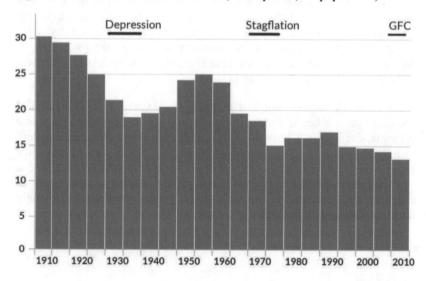

The birth rate was at its lowest during periods of economic malaise such as the 1930s, 1970s, and following the Global Financial Crisis of 2007–8. It is common to blame these birth rate "troughs" on economic recession but, as can be seen from the data, the greatest declines in birth rate occurred in the periods of prosperity that *preceded* the recessions such as the 1920s and 1960s. In each case, the birth rate had been dropping for at least fifteen years before the recession. Also, at least during the first two instances, the birth rate then began to rise *during* the recession. This eventually produced the baby boom of the 1950s, as well as slight rises in the late 1970s and 1980s.

The character of the population also differed between these periods of falling and rising birth rates. When the birth rate was falling, such as in the 1920s and 1960s, behavior tended to be hedonistic and pleasure-seeking. Skirts became shorter and sexual mores looser. There was also, almost by definition, a reduced interest in having children. All of this indicates a fall in C. Or, because C was falling throughout the late nineteenth and early twentieth centuries, these were periods of an *especially rapid* fall in C.

The late 1940s and early 1950s were more conservative and, by definition, people had more interest in children. These were times when the fall in C slowed, or even reversed slightly. Tougher times may also have increased the level of child V. This also applies to the 1840s and 1850s, when birth rates were rising in England, and society was at its most puritanical. In this case there was a substantial rise in C.

Biohistory helps explain why these fluctuations take place, and in particular why major recessions occur, but also why political instability is associated with them. What we are seeing is another cyclical pattern. Because recessions and political instability are so characteristic of the "trough" years, we refer to this pattern as the *recession cycle*.

According to the recession cycle hypothesis, recessions take place because in previous years there has already been a sustained rise in the level of stress in the general population, and especially the influential younger generation. In other words, there is a society-wide stress reaction, involving rapidly rising levels of stress hormones, which eventually engenders a pervasive sense of fear and loss of confidence. This in turn causes people to invest more cautiously in the market—or stop investing altogether—which causes the market to crash.

So what causes stress levels to rise so high to begin with? To understand this we need to look at two key factors that determine the level of stress, as discussed in earlier chapters.

The first is the overall level of V in the population. High V helps populations to migrate by making them more confident, aggressive and cooperative. It also drives them to migrate by reducing their tolerance for crowding. Thus, *at any given level of population density*, they experience greater stress than lower-V populations.

The second main factor is the level of C. This does not so much affect the *level* of population density but the way it is perceived, based on how much other people are sought out or avoided. High-C peoples tend to be hardworking, introverted, and religious, and to have lower testosterone. They also tend to avoid stimulation, which lowers their levels of stress. Lower C peoples are more hedonistic and pleasure seeking. They tend to have wide social networks and enjoy crowded parties and exciting and competitive occupations. This cascade of high-density experience induces a rise in stress.

To illustrate how this works, imagine two people, John and Brad, who have the same levels of V, which means their intolerance of crowding is much the same. If they were both placed in the same environment, each seeing the same number of people per day, they would both experience the same level of stress. However, John is high C, which means he prefers the quiet life. He is married, spends time with his family, or with a small circle of friends. He is an accountant, goes to church and is happy with his life. Brad on the other hand has lower C. He is a trader in a big securities firm, a high-pressure job with big bonuses but also big losses. He has separated from his wife and has a stormy relationship with a co-worker. He goes to nightclubs two to three times a week and has a huge circle of friends.

Both men live in the same area, but their environments are completely different. Based on the number of people he sees and the background noise, John might as well live in quiet rural location. Brad, on the other hand, lives the experience of a big and noisy city. Although the actual population is similar for both, Brad's *perception* is of a much higher population density.

The same applies to other forms of stress. John, with higher C and thus lower testosterone, prefers quiet recreations such as family picnics or mixing with a small group of friends. Brad, with lower C and thus higher testosterone, prefers hunting, bungee jumping and drag racing.

It might seem strange that people would choose activities that increase stress, but this is a well-known facet of human psychology. A common reaction to excitement, novelty and danger is a rise in adrenaline and cortisol, both stress hormones. At moderate levels this can be pleasant, even euphoric. It is exactly the thrill that people get from extreme sports. The danger is what makes it attractive. The same can be said of the crowds, flashing lights and loud, thumping music at a nightclub, as well as the thrill experienced by elite traders in a fast-moving market during a boom. But the fact that a particular kind of stress may be welcome doesn't make it safe. People who experienced divorce, the death of a spouse or financial crisis suffer an increased risk of illness and death. Surprisingly, positive events such as marriage, a change of job, the birth of a child or outstanding personal success are also stressors which increase the chance of illness or death.[204]

Given that both Brad and John have equal levels of V, Brad's level of stress would tend to rise and John's to fall. If there are more people in the society with high C like John, the general level of stress will fall. If there are more people with lower C like Brad, the general level of stress will rise.

The other factor to be aware of is that levels of C, V and stress tend to equalize themselves across populations. People in the same nation tend to share lemming and civilization cycles, though there may be regional differences, as seen in the origins of the American Civil War (when the South's G year was about twenty years after that of the North). Even neighboring nations such as England, France and Germany tend to have cycles of similar length but with the G periods a few decades apart. In other words, the closer the relationship between peoples, the more similar they will tend to be in these terms.

How this equalization takes place is a matter for conjecture. Shared cultural experiences undoubtedly have some influence in human societies. For example, if bungee jumping is popular because of the prevalence of high-testosterone people, then even those with lower testosterone are more likely to give it a try. But it is very likely that pheromones play a role. The reason for proposing this is that lemming cycles in animal populations show the same pattern of local synchronization as found in humans. Lemmings and muskrats do not read newspapers or watch TV, so some other factor must be involved. Some of our recent research, as yet unpublished for commercial reasons, suggests that pheromones can have a powerful effect on C, V and stress. This implies that if more people in a society act to raise their levels of stress, the general level of stress in the society will tend to rise—even in groups who are not bungee jumping or frequenting discos.

The effects of recession

The question answer now is *why* the level of C should fall faster in some periods than in others. Why did people in the 1920s and 1960s have lower C than in other decades?

The first point to recognize is that the *population as a whole* need not experience a major change in C so long as there is *enough change in the most influential age cohort*. We saw in the last chapter that nations tend to plunge into war when a super-aggressive generation reaches its peak military age, especially in and around the early twenties. In modern Western societies the most influential generation tends to be the youth generation, which sets the trends and affects the national mood. Anyone who lived through the 1960s can remember how miniskirts and sexual liberation dominated the cultural landscape, whatever the objections of the older generation.

A person's level of C can change at any time during the course of their life, but childhood and early adolescence are by far the most important in setting the average lifetime level of C. Thus, people growing up in tougher times such as the 1890s and 1930s tend to have higher C than those maturing in times of prosperity. Based on animal experiments which suggest that C-promoters in infancy do not affect or may even increase testosterone (see chapter two), the implication is that people who experience economic recession from the age of about 5 should have generally higher C and thus lower testosterone.

To be more precise, the Great Depression would have most strongly influenced the C of the age cohort born in 1925, who experienced it in both late childhood and adolescence. Members of this cohort came of age in 1946 and passed their early twenties in 1945–50. Together with those born somewhat earlier or later, this was the stable and conservative generation which gave rise to the baby boom. It is no coincidence that the peak of the birth rate in the mid-1950s coincides exactly with this highest

C generation reaching the age of 30, which is the core of the childrearing years.

Margaret Mead makes exactly the same observation of the Manus people. Their traditional childrearing system involved tight control of infants but very little control of older children or adolescents, thus creating the uniquely changeable Manus character. But, as indicated in chapter five, the generation experiencing late childhood and late adolescence in the difficult war years were far more conservative in their thinking. The Japanese occupation was also highly stressful, which would have increased their child V.

The return of prosperity

By 1940, aided by the stimulus of World War II, the Great Depression was over. After 1945 the return of peace led to an unprecedented era of prosperity. Cars, home ownership and TVs spread rapidly through the population. This prosperity continued until the close of the 1960s, after which the "stagflation" of the 1970s marked more uncertain times. The cohort whose temperament was most shaped by prosperity would be that which passed its late childhood and adolescence wholly within this period, which is essentially people born in the 1940s. Together with those born somewhat before or after, this was the youth generation of the 1960s which pursued sexual liberation, drug experimentation and other rejections of traditional social norms as summarized by Timothy Leary in 1966: "Turn on, tune in, drop out."

This can all be explained as a result of prosperity lowering C and thus causing a surge in testosterone. It is the same behavior that characterized the youth generation of the 1920s, who had also grown up in the prosperity that followed the hard times of the 1890s. In both cases it was associated with a falling birth rate. It is again no coincidence that the lowest level of birth rate, in the mid-1970s, coincides exactly with the lowest C generation born around 1945.

The crash

Hard times experienced in late childhood and adolescence increase C in the influential youth generation, causing changes in behavior that *reduce* stress. By contrast, experience of prosperity in late childhood and adolescence reduces C, causing changes in behavior that *increase* the overall level of stress. If we put these two factors together, we begin to see an explanation for the crash of 1929 and the depression that followed.

To some degree, a rise in the level of stress is no impediment to prosperity. High testosterone is associated with confidence, optimism and energy. People with those characteristics tend to be gamblers, ever ready to believe that prices will keep going higher. The stock boom of the 1920s reflected this attitude, which may even have been enhanced by a small rise in stress since anxiety has been linked to problem gambling and poor judgment.[205] And a rise in the level of stress is likely to increase the level of anxiety in segments of the population, though not necessarily among the bungee-jumping extroverts described earlier.

But while a moderate increase in stress is compatible with optimism, greater levels bring fear. When enough individuals have crossed that line and the markets begin to fall, this intensifies the level of stress, which in turn leads to a cascading stress reaction that causes people to become overly cautious and pessimistic. The result is a surge in anxiety and a collapse of the stock market, the property market, or whatever has been driving the boom. And unlike market crashes such as 1921 which quickly correct themselves, they lead to a prolonged malaise.

The crises that followed 1929 and 2007 can be seen as society-wide stress attacks, the culmination of rising stress levels during the period of prosperity. No government policy could contain them because they occurred on a physiological level. We are accustomed to viewing exuberant prosperity as the opposite of anxiety-filled recessions, but if the C hypothesis is correct prosperity itself is the *cause* of the recession. The irrational optimism before the crash and the irrational pessimism after it are two sides of the same coin.

Putting all this together provides an overall picture of the twentieth and early twenty-first centuries in terms of C and V. The picture is especially applicable to America but has relevance for other Western countries. Its basis is a prolonged decline in both V and C as a consequence of the civilization cycle. The fall of V has acted to reduce sensitivity to stress while the fall of C has attracted people to stress-promoting activities, whether bungee jumping or intense partying. If both declined by a similar amount then the net effect on stress would be zero, since greater exposure to stimulus balances a greater tolerance of it.

But there are eras when the level of stress rises much faster, as a result of a generation growing up in prosperity and thus entering adult life with unusually low C. In this case their search for stimulus outpaces their tolerance for it, so stress rises. Another generation grows up in hard times so that their level of C is relatively high compared to their tolerance, so stress falls. This gives rise to fluctuations in the level of stress as described in Fig. 10.2 below.

Fig. 10.2. Uneven decline in C as an explanation for surges in stress levels. As C drops faster than V in hedonistic eras such as the 1920s and 1960s, stress rises causing recession and (in the 1930s) political terror.

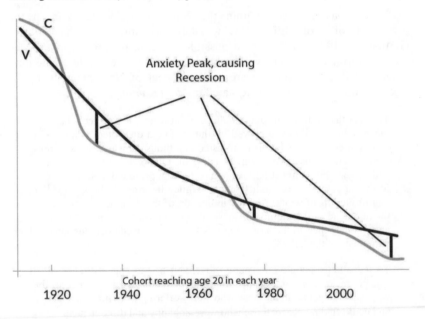

Political terror

Economic recessions can lead to serious personal and financial hardship, but there are even more grave effects when C falls faster than V. High levels of stress are deeply unpleasant, and people experiencing them tend to be angry or fearful, and suspect others of ill will. We have seen how the highly stressed Mundugumor were filled with anger and mutual suspicion. Similarly, highly anxious Germans in the 1930s were more likely to join the Nazi Party, which preached hostility and hatred towards Jews and other minority groups. Young men were given free license to vent their anger by attacking political rivals and smashing shop windows, and by 1934 Nazi thugs were beating up strike-breakers and passers-by. Historian Richard Evans describes the innate aggression of the SA, the Nazis' private army, after the party came to power:

> As the young brownshirts found their violent energies deprived of an overtly political outlet, they became involved in increasing numbers of brawls and fights all over Germany, often without any obvious political motive. Gangs of storm troopers got drunk, caused disturbances late at night, beat up innocent passers-by, and attacked the police if they tried to stop them.[206]

Insecurity was also rife within the Nazi hierarchy, as seen in the personality of Adolf Hitler. Having established himself as Chancellor of Germany in 1933 and sensing a possible threat to his power, in 1934 he ordered various potential rivals to be dispatched in the "night of the long knives." Hitler was quite aware of the appeal of Nazi ideology to the anxious and discontented, as he stated in *Mein Kampf*:

> The fact that millions of our people yearn at heart for a radical change in our present conditions is proved by the profound discontent which exists among them. This feeling is manifested in a thousand ways. Some express it in a form of discouragement and despair. Others show it in resentment and anger and indignation. Among some the profound discontent calls forth an attitude of indifference, while it urges others to violent manifestations of wrath. Another indication of this feeling may be seen on the one hand in the attitude of those who abstain from voting at elections and, on the other, in the large numbers of those who side with the fanatical extremists of the left wing.

> To these latter people our young movement had to appeal first of all. It was not meant to be an organization for contented and satisfied people, but was meant to gather in all those who were suffering from profound anxiety and could find no peace, those who were unhappy and discontented.[207]

Hitler was aware that anxious young people were also attracted to Communism, which flourished at the same time, especially in its more violent forms. Though ideologically opposed to one another, Communism and Nazism were psychologically similar in that both legitimized hostility and anger. Both were therefore more readily accepted by people who were anxious and highly stressed in the first place.

This confrontation was played out most fully in Spain in the 1930s, where both Communism and Fascism flourished, bringing about the civil war of

1936–39. In their brutality and commitment to government control of the economy, the victorious Fascists had much in common with their opponents, differing most markedly in their attitude towards religion. Once again, the explanation for the appeal of both Communism and Fascism in Spain is that Spaniards also experienced a surge of stress in the 1930s, along with the rest of Europe and North America.[208]

The same analysis can be applied to Russia, with the major difference being that the Russians were already highly stressed. Writing in the late 1890s, Olga Semyonova Tian-Shanskaia gives us a vivid account of the treatment of children among the Russian peasantry:

> Up to the time Ivan takes his first steps, he is looked after by his sister, a girl of nine or ten years of age. She has difficultly carrying him and drops him … Sometimes Ivan tumbles down a hillock. When he cries, his baby-sister uses her free hand to slap him on the face or head, saying "keep quiet, you son of a bitch." Sometimes his sister leaves him on the ground … For an hour or more, the child crawls around in the mud, wet, covered with dirt, and crying.[209]

Infant deaths were also common in this period, and some were thought to be intentional; the prevalence of infanticide is strong evidence of a highly stressed society. Children were also treated harshly, beaten for such offenses as screaming, getting muddy, stealing food, lying, swearing and fighting. Domestic violence was endemic, with husbands often severely beating their wives, sometimes to death.[210]

All these indications paint a picture of Russia in the late nineteenth century as far more stressed than Western Europe or America. Like them, Russia too was experiencing a decline in C, as observed by loosening sexual morals.[211] This decline in C and subsequent rise in testosterone were strongest in the cities, and so it is no surprise that Bolshevik support was strongest in urban areas.

When the Tsar was overthrown in February 1917 and replaced by a democratic socialist provisional government, the stressed and anxious Russians did not accept the change because they were psychologically primed to accept a more brutal and authoritarian form of government. Accordingly, a few months later the October Revolution swept away the liberal Mensheviks and brought the brutal Bolsheviks to power.

It is because of the rise in stress that countries like Germany and Russia turned to brutal authoritarianism, despite having had their monarchies replaced by relatively liberal democratic systems like Russia's provisional government and Germany's Weimar republic.

A bizarre but commonly expressed desire among workers and peasants during the 1917 revolution was for a "democratic" government but with a "strong Tsar" at the head of it.[212] This does not mean they would have voted for dictatorship, if given the chance, but they were more ready to accept its dictates once in power. Thus it is that when Russia shared in a European surge of stress during the 1930s, it turned from an already brutal autocracy to a far more brutal and even murderous one with show trials, gulags and mass murders. Fear and suspicion were rampant at all levels of society, making people ready to accept accusations that this or that person was a traitor or saboteur. One historian describes the effects of this stress at its height:

> The "Gulag archipelago" of labour camps strung across the less hospitable parts of the Soviet Union, above all in Siberia, swelled to bursting with millions of prisoners by the late 1930s. From Stalin's acquisition of supreme power at the end of the 1920s to his death in 1953, it has been estimated that over three quarters of a million people were executed in the Soviet Union, while at least two and three-quarter million died in the camps. In this atmosphere of terror, fear and mutual recrimination, anything out of the ordinary could become the pretext for arrest, imprisonment, torture and execution.[213]

To us, looking back, this sort of behavior makes no sense, but then we are not psychologically attuned to see treachery around every corner.

Nor are we primed to submit to brutal authority. But it was the very brutality of the government which caused the highly anxious Russians of that time to obey it, while the same brutality would cause less anxious people to resist it. As discussed earlier with reference to the Middle East, government reflects the prevailing attitudes of the population even when most people hate it. Anxious people tend to accept and obey powerful authority, whether elected or imposed. Stressed Americans in the 1930s accepted a more powerful government under the New Deal than they had supported in the 1920s. Stressed Russians in the 1930s also accepted a more powerful government than they had in the 1920s. Stalin was more powerful and brutal than Roosevelt because Russians were *already* more anxious than Americans, so the same rise in stress took this to a much higher level. The Russian Terror of the 1930s was not a product of one murderous dictator, but of a whole society driven mad by stress.

Human societies under stress behave in ways that are fundamentally similar to the Whipsnade baboons described in chapter four. Under extreme stress the normally protective dominant male began to attack other baboons without provocation. Likewise, each animal below him was liable to attacking those beneath, to the extent that almost all females and young were killed. In other words, the autocrat does not create the stress, the stress creates the autocrat.

Thus, Stalin was the product of stress rather than the cause of it. And as stress ebbed during the 1940s and early 1950s so too did the terror, with no change of regime required.

Stressed animals and people tend to give higher status individuals more power. In humans this can translate into governments taking or being granted more power over the economy. The people liable to do this the most are of course the Communists, whose policies cost millions of lives while holding back economic growth. Mao's "Great Leap Forward" of the 1950s was intended to make China into an industrial power, but instead caused a horrific famine.

Fascist governments interfered quite heavily in the economy, as did Roosevelt in the US. His administration acted to raise the price of food, fix minimum wages, regulate and insure bank deposits, and implement a huge program of public works and unemployment relief. Much of this would be uncontroversial today but it was a huge expansion of federal power for the time.

The relative strength and brutality of a government depends on the absolute level of stress. The governments of the United States, Germany and Russia all became more autocratic during the 1930s as stress increased (see Fig. 10.3 below). The percentage rise in stress was probably similar in all three cases, but the effect on each society was different because each started with a different level of stress.

The End of Terror

The Depression and turmoil of the 1930s meant that a generation grew up with higher C. This, plus falling V, meant stress levels eventually receded. Americans became the sober, family-loving parents of the baby boom, and Germans became far less willing to accept authoritarian rule than they had in the 1930s, thus making an easy transition to liberal democracy. Stalinist terror died away in Russia, leading to a far more moderate regime even in Stalin's lifetime.

Fig. 10.3. 1930s surge of stress in Russia, Germany and the US. Rising stress causes society to become more autocratic and brutal. The level of autocracy and brutality depends on the initial level of stress.

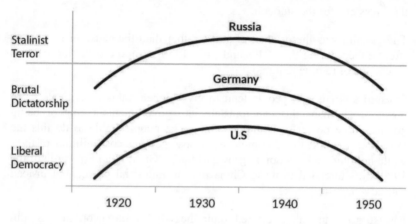

In the West, stress rose again in the late 1960s but not nearly as much as in the 1930s. Thus, instead of global recession and political terror came mass student rebellion. There were some violent confrontation, but nothing like those of the 1930s. In recent years, declining stress in Russia has made a transition to democracy possible, though of a peculiarly autocratic form since Russians are still more inclined to autocratic rule than other Europeans.

The Global Financial Crisis

The recurring theme of this chapter is that governments have little or no power to halt the underlying forces of economic and political change, because these forces are driven by changes in temperament. Governments could not prevent or cure the Great Depression, and could not halt the rise of dictatorships. And in 2007 they could neither prevent nor cure the global financial crisis.

By the end of the twentieth century a generation came of age that had known nothing but prosperity. Once again, as in the 1920s, there was a speculative boom, but this time in property rather than stocks. People borrowed more than they could afford to buy bigger houses than they needed, with the idea that ever-rising prices would make them rich. Governments supported this by arranging loans to marginal borrowers through government-backed institutions such as the US Federal National Mortgage Association ("Fannie Mae") and Federal Home Loan Mortgage Corporation ("Freddie Mack"). Trading houses bought up billions in mortgages without adequately checking the security behind them, again with the assumption of ever-rising prices.

This environment can itself be productive of stress. In *The Hour Between Dog and Wolf* (2012), John Coates gives a vivid picture of traders in the financial markets in the period leading up to the Global Financial Crisis.[214] This fast-paced business requires split-second decision-making and an appetite for risk, and is thus one of the few professions that benefit from high testosterone. Success can bring enormous prestige and multi-million dollar bonuses, but such an environment also tends to increase stress.

In this environment, once again stress rose past the tipping point and the market crashed. Coates' study shows how this impacted on the market traders. As the market plunged, traders' testosterone fell and cortisol rose, creating problems opposite to those of the boom. Stressed traders stopped being irrationally optimistic and became so risk-averse that they failed to see clear opportunities for gain. Buying slumped and the crunch was even worse than it should have been. Even a drastic cut in interest rates by the US Federal Reserve had no great impact, because the problem was hormonal and psychological rather than rational.

Biohistory suggests that the general population experienced the same hormonal and psychological changes, also driven by stimulus-seeking behavior as a result of falling C. And, once again, it proposed that the key rise in stress was that which happened *before* the crash.

From this viewpoint an economic recession can be seen as a kind of nationwide—or even worldwide—stress attack. And that is why government actions, no matter how soundly guided by economic principles, have so little effect. The basic problem is not economic or even cultural but physiological.

Recovery

But this recession has had its own impact, very similar to that of the 1930s though on a lesser scale. The American birth rate has in recent years risen slightly, as indicated in Fig. 10.1 above. More than that, there has been a significant change in the character of the youth generation passing late childhood and adolescence since the GFC. Teenagers in particular are less likely to drink alcohol, commit crimes, smoke and use recreational drugs. Sexually transmitted disease and teenage pregnancies are down. They are less likely to visit bars and cafes, causing many such businesses in Europe to shut down. There is even evidence that they are more polite.[215] All this is exactly what should be expected from a rise in C, and it will of course be reversed as the effect of the downturn fades and C continues to fall.

Thus it is that many of the changes taking place in the West over the past century can be explained in terms of one simple notion. People who spend their late childhood and adolescence in times of prosperity have lower C and lower child V. This makes them unconventional, pleasure-seeking, extroverted, and likely to seek out stimulus beyond their underlying tolerance for it. They thus experience a rise in stress which, applying especially to the influential youth generation, causes recession.

People who spend their late childhood and adolescence during the recession have higher C and higher child V. This makes them more conventional, introverted, and likely to avoid stimulus. They thus experience a fall in stress which, as they become the influential youth generation in turn, ends the economic malaise. Unfortunately, as will be discussed in chapter sixteen, the continuing fall in C means there is no guaranteed return to prosperity.

Testing

Long-term measurement of stress hormones, such as salivary cortisol, would test the prediction that stress hormones rise steadily *before* a recession starts and begin to decline before it is over. The sample would need to be broadly representative of the population and especially the young. In principle, it should be possible to anticipate recessions before they happen.

In addition, age cohorts which pass their late childhood and adolescence during times of prosperity should have higher testosterone and other signs of lower C. Those who pass their late childhood and adolescence during times of recession should have lower testosterone and other signs of higher C.

CHAPTER ELEVEN

WHY REGIMES FALL
AND CIVILIZATIONS COLLAPSE

The nature of peoples is first crude, then severe, then benign, then delicate, finally dissolute.
—Giambattista Vico

In earlier chapters we have seen that high levels of C and V are essential to the survival and success of a civilization. C promotes hard work, business skills and obedience to the law. Child V promotes acceptance of authority, making states unified and cohesive, and V gives them the warlike aggression necessary to win in battle.

In this chapter we will see what happens when V and C decline, mainly as a result of urbanism and prosperity. In pre-literate peoples the onset of prosperity often leads to an increase in aggression as plentiful food increases population and thus V, while declining C causes testosterone to surge. But in civilized societies the effects of prosperity are uniformly disastrous in the medium to long-term. C-promoting and V-promoting behaviors are difficult to maintain in any circumstances, and become far more so when prosperity and urban life act to reduce them.

Elites are especially vulnerable because they have plentiful food and tend to live in cities, and so a common pattern is for them to be replaced by new and more vigorous (generally higher V) groups. Popular Revolutions such as those of Cuba and China represent just a replacement, as does the conquest of ancient civilizations by nomads and other "barbarian" invaders.

In other cases the decline of C and V in the general population is so extreme that the original population is largely absorbed, expelled, or even wiped out.

The Decline of Athens

In the late fifth century BC the Persian Empire was the biggest the world
had ever seen, and at the height of its power and influence. It ruled over a
multitude of peoples from Egypt to India, had enormous wealth and an
almost inexhaustible supply of men. In 480 BC the Emperor Xerxes I sent
a vast army and fleet to subdue the troublesome city states of Greece. The
Greeks resisted with bitter determination.

At the pass of Thermopylae, three hundred Spartans and their allies were
slaughtered to the last man rather than submit. The people of Athens fled
across the sea to the island of Salamis. Destitute and in exile, they watched
the Persians ravage their land and burn their city to the ground. Xerxes
offered the most generous terms. If they would join him as an ally they
would pay no tribute, and he would make them the leading city in Greece.
They refused and vowed to fight him "as long as the sun moves in the
sky."

Against great odds the Persians were defeated, first by sea at Salamis and
later by land at the battle of Plataea. During the decades that followed,
democratic Athens saw an explosion of creative energy, achieving new
heights in literature, philosophy, art and architecture that have never been
equalled.

Then in 338 BC the Macedonian King Phillip again threatened the
freedom of Greece. After a single major victory, at the battle of
Chaeronea, he offered the Athenians the same terms as had Xerxes 150
years earlier. This time they submitted with barely a whimper. The
Macedonians and their Greek subjects went on to subdue the Persian
Empire, but Greece itself was no more than a backwater in these new
Empires. Later the Greeks were to give way even more abjectly to Rome,
and they continued to decline. Independence and democracy lost,
commercially feeble and militarily weak, Greece was the merest shadow
of its former glory. Eventually, the Roman Empire itself collapsed into
barbaric anarchy.

Greece and Rome are far from being the only civilizations to have fallen,
so what causes this recurring pattern? So far we have considered the
cyclical patterns which cause civilizations to rise and evolve, and to
splinter and reform and make war. We have also looked at the forces that
can cause a civilization to decline. In this chapter we will settle the
question of how and why they fall.

Regime change and civilization collapse

Political systems are inherently unstable. Even when a civilization is rising, the individual regimes and dynasties that rule it rarely last more than a few centuries before falling or being overrun by incomers, as occurred several times in Chinese imperial history.

A major factor in this pattern is the turmoil of the G-90 phase of the lemming cycle, when population growth is beginning to rise from the point of lowest G. At this time, infant C is high and stress low, so that political loyalties become more local. The typical result is chaos. No Chinese dynasty has ever survived the G-90 period, including the most recent one (1920) when the Qing dynasty collapsed into warlord anarchy, followed by a civil war and Communist victory. Likewise in England, the G-90 period around 1460 saw the ruling house replaced not once but three times.

At times of high C regimes can also be vulnerable during the period after the G year, when V is at a peak. At such times people with strong impersonal loyalties (from high C) may have the energy to overthrow rulers who no longer reflect their interests. The French and Russian Revolutions are examples of this.[216]

Invasion by a powerful enemy can cause regimes to collapse without destroying the civilization. Lemming cycles have an impact on this process too. A society will be far better able to resist foreign invasion when its V is high and its leadership strong during the periods close to the G year (though a sufficiently powerful enemy could still depose any existing regime). Conversely, a society in the G-90 period will be much more susceptible to foreign invasion. For example, it was the turmoil associated with a G-90 period that allowed the Manchus to conquer China in the early seventeenth century.

There is another, broader principle behind regime collapse, which is that groups with higher V (and to a lesser extent high C) tend to overcome groups with lower V. This applies not only to the conquest of Athens by Macedon and of China by the Manchus, but to the success of Communist revolutionaries such as Mao and Castro, and to the decline of wealthy urban aristocracies. It also explains some curious aspects of the behavior of pre-literate peoples. This chapter will explore all these scenarios.

Reasons for the loss of C and V

There are several reasons for the collapse of C and V, but the most important is prosperity. Over thousands of years societies have raised their C and V by adopting religious and cultural practices that give them an advantage over their competitors. But no matter how strong the cultural system, it is difficult to maintain C and V in the long-term without at least occasional food shortages. And successful civilizations tend to become affluent.

The raising of C and V over thousands of years might be compared to a band of men, with great effort, pushing a heavy boulder up a hill inch by inch. The effort required is close to the limit of their strength, but the fact that their very survival depends on it forces them on. Then a stranger walks down the hill towards them and starts pushing the boulder downhill. It slows to a halt and starts to slip backward, slowly at first and then faster as the team's feet scrabble for purchase. That stranger is prosperity.

Population density also affects V. In general, V is increased by occasional stresses such as famine or external threats which activate the stress reaction but then allow it time to recover. This is the "toughening" effect referred to in chapter four. But *chronic* stress undermines V because it allows no time for recovery, and also because it may cause mothers to neglect or abuse their infants. For people with high V, who are intolerant of crowding, the urban environment of a developed civilization is a source of chronic stress. Thus, population density as such tends to undermine V, all the more when it is combined with prosperity.

Prosperity and aggression among pre-literate peoples

A recurring theme throughout history is that civilized societies are overrun by barbarians from marginal lands, who are more aggressive than settled populations. But when the comparison is between hunter-gatherer and small-scale agricultural societies, the pattern is entirely opposite. Peoples on fertile plains are typically *more* aggressive and warlike than those in marginal environments. The Yanomamo of the Amazon illustrate this principle, the groups in the precarious highlands being distinctly less warlike than those on the fertile lowlands.[217] The same can be said of Margaret Mead's Highland Arapesh in their poverty-stricken uplands, who were far less aggressive than their more affluent relatives the Plains Arapesh.

Why should this be so? A clue can be found in the behavior and lifestyle of the Lakota Sioux. During the Indian wars of the mid- to late-nineteenth century, the Sioux were the largest and one of the most warlike tribes in North America. But this was a relatively recent development. Until the early eighteenth century they had been a small band of hunters, gathering wild foods and growing small plots of corn, squash and tobacco. Forced west to the Missouri by more powerful tribes they gained access to horses and guns, and when smallpox wiped out most of the rival tribes they were able to take up bison hunting on the Great Plains. This was so rewarding as a source of food that they abandoned agriculture altogether.[218]

But instead of V declining with access to plentiful food, it *increased*. The status of women fell and the Sioux became ferociously warlike. During this period they attacked rival tribes mercilessly, either killing them or driving them from their lands. In 1873, a thousand Sioux warriors killed nearly two hundred Pawnees, men, women and children, for trespassing and hunting. There was no peacetime for the Sioux.[219] Along with their violence went a supreme self-confidence, all indications of high V.[220]

As we seen in previous chapters, if people are less anxious as adults than as children, they have higher V. In addition, once any group of previously marginalized peoples becomes prosperous, C decreases and testosterone rises, leading to a surge of aggression. Both of these factors were used in chapter nine to explain the World Wars of the early twentieth century. Thus, we should *expect* any previously marginal group that encounters abundant food to become more aggressive. As more children survive and neighboring peoples are killed or expelled, population expands quickly which combines with increased patriarchy to reinforce V for a time. Thus it is that the *short-term* response to more plentiful food is higher V.

But this does not last long. Among pre-literate peoples whose cultures lack strong V-promoters, the initial surge of V drops rapidly after a few generations, allowing other groups to move in and take over. The expansions of the Yanomamo and the Sioux were terminated by colonial rule, but biohistory indicates their expansion would have faded with continuing prosperity.

When pre-literate peoples are not displaced in this way, the full effects of declining V may be observed. The Tchambuli had settled in a fertile lakeside area which supported a relatively dense population. They seem at one time to have been fierce head-hunters, as indicated by the requirement that every boy kill a man for ceremonial reasons, but at the time of

Western contact they were distinctly low in V. Men were so unaggressive that captives for sacrifice were sometimes purchased rather than captured in war. Some twelve years before contact they had been driven from their home by more warlike peoples.[221]

In most circumstances the changeover happens far more quickly than this. Pre-literate peoples in lowland areas have no great advantage in technology, numbers or size of political units, so any loss of V would quickly cause their displacement. Thus it is that the lowlanders are more aggressive than peoples from the marginal lands because as soon as they lose V they are driven out, and the invaders gain a temporary surge of V from suddenly plentiful food.

Prosperity and aggression in civilization

In one sense, nothing changes when civilizations arise. Peoples in more fertile areas still tend to lose V since they are less subject to famine than those in the marginal lands. But their displacement by higher V peoples now happens far more slowly. V tends to find support in the development of effective V-promoters, commonly tied to religious systems, so it declines more slowly. Also, farming populations are denser and (given higher C) can form larger political units. Technologies such as metal-working, defensive walls and writing provide further advantages. Thus it is that the V of settled peoples can drop below that of the "barbarians" without them being immediately overrun.

There are certain groups, however, which are especially liable to loss of V. Ruling elites tend to be better fed and live more comfortable lives, protected from wind and weather. Even when men of the ruling classes are warriors, their wives live in relative comfort. The other factor is city life with its high population densities. The strongest effects are found when both these factors coincide, as it does when ruling elite live in cities. The loss of V also entails an eventual loss of C, with its undermining of wider loyalties and the principles upon which social cohesion depends.

If it is only the V and C of the elite that is undermined, higher V invaders tend to displace the elite and form a new ruling class. This new elite becomes wealthy in its turn so that its own V and then C are undermined. But if the decline in V and C is general throughout the entire society, the whole population may be overrun, causing either the destruction of the entire civilization or at least the supplanting of swathes of its people.

After the Norman Conquest of England in 1066, the native English kept their language and identity largely intact and eventually absorbed the Norman incomers. By contrast, the Anglo-Saxons who colonized Britain six-hundred years earlier virtually obliterated not only the native language and culture of Romano-Celtic Britain but even the population's Y chromosome outside the borders of Wales.[222] The long Roman peace had so sapped the V of the native population that invaders easily overcame their resistance. Fig. 11.1 below helps to put this into perspective.

Fig. 11.1. Declining C and V cause regimes to collapse or be overrun. The greater the decline, the more serious the collapse.

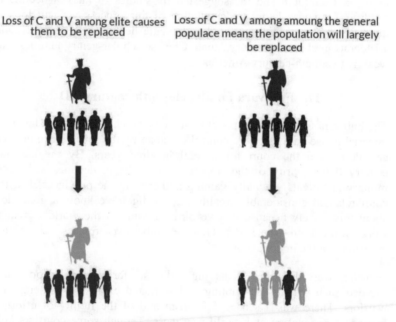

The conquest of settled peoples by mobile, warlike tribes from harsher environments is one of the great universals of history. We see it in the Middle East, in Europe, in India and China, in Central America and Peru.

But of course regimes eventually collapse even when there are no obvious barbarians to do the work, as with the decline of the Later Han dynasty in China in the second century AD. Much has been written of the "mysterious" collapse of the Mayan civilization in Guatemala. Explanations have been offered which include natural disasters, intensified warfare, famine due to over-population, and civil strife.[223] But once we understand the nature of C and V, there is nothing mysterious about it. The key factor is prosperity, so that the very success of a civilization is what destroys it.

Declining V and C in wealthy urban elites

In an elite group undergoing a decline in C and V, there are certain things we should expect to see. Low V should cause reductions in energy, confidence and morale. From declining C we should see a loss of the work ethic, of self-control and adherence to rules of behavior, of interest in family and children and (as suggested by studies of rats), a decline in fertility. To gain a better understanding of this process we will look at three urban elites which suffered a dramatic decline in V and C: court aristocrats in eleventh century Japan, Chinese scholar-gentry families, and wealthy twentieth-century Americans.

The Fujiwara family, eleventh century AD

The Fujiwara were descendants of aristocrats who had helped the Japanese imperial house gain effective control of Japan by the seventh century AD and dominated the court for several hundred years. By the eleventh century the authority of the imperial court was in decline, with local military aristocrats gradually gaining authority in the provinces. But the court retained considerable wealth and prestige. We know a great deal about this society because it gave birth to some of the world's greatest works of literature—the Tale of Genji and other masterpieces were written by women in the imperial court.[224]

Courtiers were described as having a distaste for physical effort. They avoided such pursuits as hunting and war, and expressed contempt for warriors. There was a sense of fatalism and of the futility of effort; a feeling of resignation and world-weariness. Though some courtiers held jobs, many had no interest in work and few had serious intellectual interests. Sexual behavior was notably free, to the extent that child paternity was often doubtful. Unattached women in particular were highly promiscuous. Women had little interest in children, though for political reasons they seem to have borne many. Consistent with low V is the fact that women held relatively high status in the court, higher than at any time in Japanese history prior to 1945. They could inherit and keep property, and unmarried women had a great deal of freedom.[225]

As with the Tchambuli, this was a society obsessed with artistic expression. There was an intense focus on poetry, on the details of dress and personal appearance, and a deep sense of appreciation for natural beauty which was to have a marked influence on future generations of Japanese. It is not possible to link artistic creativity too closely to low V, since the arts have also flourished in high V societies, such as Elizabethan England or Athens at the time of Pericles. But low V societies commonly show a distinctive artistic sensibility focused on external forms and appearances.

In most situations, elites that lack martial vigor and a work ethic soon lose power. It is a testament to the peculiar Japanese reverence for hereditary status that the Fujiwara and the imperial court retained their wealth and authority until this time - but not for much longer. In the following century provincial military leaders took power and reduced the Fujiwara to ritual insignificance.

Chinese scholar-gentry

Until the close of the nineteenth century, the governing Chinese bureaucracy was drawn from members of the gentry, generally those of south China. These were people who had gained wealth originally as merchants or landowners, allowing them the leisure to study for the imperial examinations. Success in the examinations required an intense work ethic and high intelligence, and only a small percentage of candidates passed. Successful candidates could go on to become high officials, providing further opportunities to accumulate wealth.

These prominent families tended to decay after a few generations of prosperity. Successful men seem to have taken little interest in their children, presumably because of a decline in C. Their sons or grandsons tended to be far less successful, showing a lack of initiative, an attitude of resignation and a sense of the futility of effort. Declining generations were also debauched and delinquent, and infertility was a major problem.[226] These changes indicate a decline in V to a very low level, and also a loss of C.

Wealthy Americans in the twentieth century

Certain groups of wealthy young Americans receiving psychiatric treatment during the 1970s provide evidence of declining C and V.[227] The individuals concerned were from the third generations of very rich families. They had been brought up from birth with the wealth generated from businesses started by their grandparents and expanded by their parents. None of these young people held jobs or pursued hobbies or intellectual interests. Their lives tended to centre around their social set, cars and clothes. They were the so-called "beautiful people." In common with the Tchambuli and the Japanese court culture, they tended to focus on appearances.

The psychological attitudes of these wealthy young Americans were similar to the Japanese courtiers and later generations of Chinese scholar-gentry. They complained of boredom, hopelessness and emptiness. They were described as easily angered, lacking shame and embarrassment, acting on impulse, having frequent brushes with the law, and pursuing compulsive sex lives.

Their families' declining interest in children can be traced generation by generation. While the parents (the second generation) had been raised primarily by servants, *their* parents (the first generation) had maintained sufficient contact with them to be active role models. But when the second generation became parents, they were occupied with their own activities and pleasures and left childrearing to a shifting body of servants, who were frequently fired when the parents became jealous of their attachment. Thus the third generation saw little of their parents. They had a great deal of freedom and relatively little consistent discipline—all circumstances which minimize C. Few of the third generation married or had children, and those who did made poor parents.

It seems that the grandparents had relatively high C, the next generation high infant C but much lower C, and the third generation low C overall. By then, V had also fallen to an absolute minimum. It must be said, however, that not all wealthy American families follow this pattern. Some with a tradition of public service, such as the Rockefellers and Kennedys, have done considerably better.[228]

These examples show how people behave when C and V fall dramatically, and also how fast this can occur. Judging by the American example it seems to take no more than two generations of ineffective parenting for them to collapse completely. But for a ruling class to lose power in most societies the change in character need not be this extreme. Any loss of C and especially V can have serious effects.

Revolution

Very often throughout history, regime change has occurred when warlike peoples move in and take over from a weakening elite. During the past three centuries we have a slightly different phenomenon—revolutionaries, who represent higher V groups from within the society. Victories by these various revolutionary groups, who have sometimes been Islamists but are more frequently communists, normally require educated leaders. But leaders alone can do little without fierce and aggressive followers willing to organize and follow and fight for them. In other words, they need high V.

We know that V is increased by cultural factors such as patriarchy, and also by two environmental factors: generally plentiful food interrupted by occasional severe famine; and rapid population growth. Times of hunger and famine followed by a period of extreme prosperity, as among the Yanomamo and the Sioux, have the strongest effect, allowing populations to grow unusually fast.

Maximum V we occurs in a population which experienced famine in the recent past but has enjoyed some generations of growth. Both the French and Russian revolutions were driven by mass support from ill-fed sections of the population, which in the preceding decades had experienced rapid growth. In fact, both Revolutions occurred soon after a long-term peak in population growth (see Figs. 11.2 and 11.3 below).

Fig. 11.2. Peak rate of natural increase (birth rate minus death rate) and the French Revolution.[229] Revolution, like war, is more likely when a generation born at the peak of population growth reaches adulthood. Figures before 1800 are for the Paris basin only.

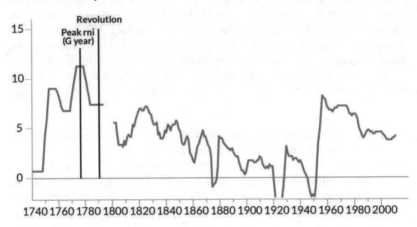

Fig. 11.3. Peak rate of natural increase (birth rate minus death rate) and the Russian Revolution.[230] The same pattern applies to the Russian Revolution of 1917 as to the French Revolution. High V people are more aggressive and more likely to overthrow their rulers.

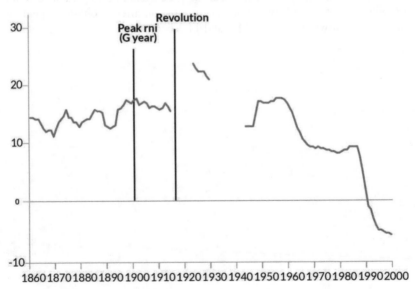

A similar pattern can be seen in the Cuban Revolution of 1957–9. Although led by urban intellectuals including the Castro brothers and Che Guevara, the insurrection was based on Oriente Province, the most mountainous area of Cuba and one with a long history of patriotic warfare. It also had an ultra-high birth rate in the 1930s. All of this suggests that Oriente had unusually high levels of V.[231] Further to this, the foot soldiers of the Revolution were the poor peasants of the rugged Sierra Maestra. In effect, the Cuban Revolution can be seen as the victory of the highest V fraction of the population, in the most mountainous area of the most rugged province in the country. As late as 1980, more than a third of the Central Committee was from Oriente.[232]

The picture is even clearer when examining the victory of the Chinese Communists in 1949. Mao's forces had been badly defeated in the centre of the country, forcing on them the devastating Long March which brought them to the Shanxi Province in the north-west, notorious for its famines and extreme climate. The famine of 1877–8, in which nine to thirteen million people died in Shanxi and neighboring areas, was one of the worst—possibly *the* worst—natural disaster in human history. This could not have directly affected the 1930s' generation, but it would give a boost to local V levels, and also made unusually rapid population growth possible. In fact, the population had still not recovered when Mao arrived. All of these factors together—the harsh climate, the recent famine, and growing population—given the people in this area exceptionally high V. Thus it was from here that the Communists fought off the Nationalists and the Japanese, and eventually emerged to take over the country.

The recent history of Iraq also makes better sense once we understand that the peoples of the 'Sunni triangle' in northern Iraq clearly have higher V than the rest of the country. This was, after all, the home base of the Assyrians and other warlike Empires that repeatedly overran the more fertile river valleys to the south. Though Sunnis are only one fifth of the Iraqi population, it was the Sunni-dominated Ba'ath Party which seized power in a 1968 coup.[233] Saddam Hussein was from this area, as were most of his key supporters. After he was overthrown, it was the centre of armed resistance to the Americans and later of ISIS, whose fierce fighters defeated much larger numbers of Iraqi government troops. These events are normally considered in terms of differences in religious ideology, but are better explained by differences in temperament.

We do not always have complete information about the background to revolution, foreign conquest or other forms of regime change, but when we do the pattern is quite consistent. Wealth and population density act strongly to undermine V and thus eventually C, leading first to military and then economic and political weakness. Famine and population growth, on the other hand, act as V-promoters which allow new people to take power.

In the next chapter we look in detail at one very significant case study of civilization collapse, showing how the concepts developed thus far can help explain how and why the Roman Empire fell.

Testing

People with behavioral indications of low V and C (low morale, poor work ethic, and poor parenting behavior) should show physiological indications of low V and C. Families resident in cities for many generations would be more likely to show such patterns, especially those with a history of affluence.

CHAPTER TWELVE

ROME

Prosperity is the measure or touchstone of virtue, for it is less difficult to bear misfortune than to remain uncorrupted by pleasure.
—Tacitus

The previous chapter showed how urbanism and prosperity undermine C and V and cause regimes and civilizations to fall. This chapter will look in more detail at the most resounding of all civilization collapses—that of the Roman Empire. A wealth of historical detail will also allow us to trace the ongoing workings of the civilization cycle, which until now has only been traced to the peak of C. Here we will see that the continuing fall in V and stress allows C to fall more and more rapidly, until at some point V ceases to fall and starts to rise. But the fall of C continues for many more centuries, leading to dissolution of the Empire and a grim Dark Age.

The history of Greece from its cultural peak in the fifth century BC is one of sad decline, but cushioned by absorption into the wealthy and cosmopolitan Roman Empire. The collapse of Rome some centuries later was far more disastrous, beginning with a devastating plague in the late second century. This was followed by civil war and chaos in the third century, a brief recovery in the early fourth century, and then shambling ruin in the fifth.

Rome, with its brutal blood sports and mass slavery, was no Garden of Eden, but the scale of this fall is hard to comprehend. Rampaging armies marched back and forth, leaving disease and starvation in their wake. Proudly literate people who had traded all over the Mediterranean were reduced to subsistence poverty. Much of the literature of the classical world was lost forever. As the population shrank, fertile fields became fever-ridden swamps.

The study of Europe and Japan showed that the rise of a civilization can be understood as a rapid rise in C, driven not only by cultural C-promoters such as sexual abstinence, discipline, and religious observance, but by a surge in V peaking around the sixteenth century. As V fell, the level of C first reached a peak (in the nineteenth century in Europe and the early twentieth century in Japan), and then began to fall. Chapter eleven showed that a decline in both V and C undermines a civilization and makes it vulnerable to collapse.

These examples to not reveal the end-point because it is some generations away, as will be explored in chapter sixteen. Rome provides an example where the entire process of rise, decline and fall plays out in a single civilization. It shows how the civilization cycle continues as C starts to fall, and how this interplays with wealth and urbanization to drive the collapse.

The rise of the Republic

The tale of Rome begins with its rise in the seventh century BC. Though a minor people largely devoted to subsistence farming, the culture from which Rome arose had been strongly influenced by earlier civilizations to the east. These civilizations provided not only technologies such as metalworking and writing but cultural technologies such as chastity and patriarchy. For example, in the early to mid-Republic Roman law gave men extraordinary power over their wives and children, and women's chastity was closely guarded, such customs serving to raise both C and V. These traditions were likely brought into Italy by the Etruscans, whose origin can be traced to Anatolia (modern Turkey) and who settled in the region north of Rome. Their influence on the local peoples was likely reinforced by Greek traders and settlers in southern Italy, whose culture was also to have an obvious and enduring impact.

This was a culture more advanced, in every sense, than the earlier one that had carried farming across Europe during the past few-thousand years. Its more powerful C and V-promoters made possible higher levels of C and V, and thus a more advanced civilization, than any previous culture in Western Europe. All of these influences were felt in the small city states of central Italy, one of which went by the name of Rome.

Rome showed signs of high C very early in its history. A republican government, which the Romans achieved in 509 BC and sustained for five centuries, requires strong impersonal loyalties because it involves loyalty to an institution or ideal more than to an individual leader. Key institutions set up at this time included the Senate and the Consulship, for which Romans had an enduring respect. They were also strongly attached to the rule of law.

The superb discipline of the Roman army, and later achievements in architecture and engineering, are further evidence of high C. This was not at the level of nineteenth-century Europe, which was driven by the still more powerful cultural technology of Christianity. Rome never industrialized and its political system was more personal in that it always emphasized patron-client relations. But Roman C was higher than anything previously seen in the West.

Early Roman history also shows signs of a rapid *increase* in C, and a civilization cycle similar to that of Europe and Japan. A peak of V and stress, as indicated by harsh punishments and political instability, can perhaps be placed in the sixth century BC, immediately before the foundation of the Republic. The regimes of this time were both brutal and unstable, with Tarquinius Priscus assassinated in 579 BC and other sixth century kings deposed and exiled. But by the end of the century Romans were better protected against arbitrary power, with a clear law code and the right to appeal to the popular assembly against a death sentence. Popularly elected tribunes had the right to veto any act of government. Greater stability and reduced inequality suggest an initial fall in V and stress.

As in Europe and Japan, this was at first accompanied by a continuing rise in C. In 509 BC the Romans threw out their king and developed more representative institutions. Significantly, the Etruscan states which covered much of northern and central Italy also became republics about this time, which suggests that the rise in C was regional rather than just confined to Rome.

Roman law was first set down in written form a half-century later, with the creation of the "Twelve Tables" in 449 BC. Among its provisos was a prohibition on making laws against individuals and severe penalties for offenses against property.[234] The purpose of the Twelve Tables was to ensure uniformity in the application of the Republic's law at the local level, and is a strong indicator of rising C.

The peak of C can be placed around 250 BC, at which point the Republic had been in existence for two and a half centuries. The population had been growing fast in the period leading up to this point, with eighteen important new Latin colonies founded between 334 and 264. Roman citizens served willingly and ably as soldiers, just as did the citizens of Europe and Japan near their peak of C in the nineteenth and early twentieth centuries. Writers of later times spoke admiringly of the "reserve" of the earlier period.

Perhaps the epitome of Roman character in this period can be seen in Cato the Censor (234–149 BC), as described by Plutarch.[235] Although he lived slightly after the peak of C, Cato was famous in his time for personifying the ideals of the earlier Republic. As a young man, before moving to Rome to pursue politics, Cato served as a soldier and gained a reputation for military excellence. In between campaigns he worked on his farm where he dressed and acted as a simple laborer, sharing duties with his slaves. He believed that farming, unlike commerce or usury, produced virtuous citizens and brave soldiers, and that the agricultural life was a consistent source of wealth and high moral values. It also required a temperament suited to consistent, routine hard work, which is a key indicator of high C.

His military aptitude and capacity for hard work were not Cato's only high-C traits. Later in his career, as governor of Sardinia, he acquired a reputation for strict integrity, reducing the costs of administration while living in austere simplicity. He administered impartial justice and severely enforced the laws against usury, banishing the guilty. Even at the height of his political career, Cato lived in an austere manner. He ate his breakfast cold, dressed simply, lived in a modest home, drank the same wine as his slaves and refused to plaster the walls of his cottages. He was regarded as an archetype of the traditional Roman character—austere, frugal, harsh, self-disciplined, honest, disinterested and impersonal when administering justice, dogmatic and a lover of order. All of this suggests that Cato, in common with the Romans of the generations preceding him, had very high C as well as C-promoting habits. While eccentric in his extremes and regarded as old fashioned by the early second century, he represented the practices and values of the early to middle Republic.

Biohistory proposes that the political character of a society reflects the temperament of its members. In effect, the Roman Republic had the character it did because until the mid-third century Romans tended to be much like Cato, though even by his time this was beginning to change.

Rising C can account for the rapid rise of Rome from a minor regional power to effective control of Italy in the third century BC, aided by a lemming cycle G period around 280 BC. A century earlier Rome had been a minor regional power, shortly to become involved in a fifty-year struggle with the mountain-dwelling Samnites of central Italy (between 343 and

290 BC). This is consistent with a G-90 period in 370 BC. By 264 BC Italy south of the Po was united under Roman leadership, resilient and aggressive enough to defeat the mighty forces of Carthage, the greatest danger it ever faced. This rapid change from local disorder to strength and unity is characteristic of the decades between G-90 and G periods.

From rising to falling C—the decline of the Republic

When Roman C peaked around the mid-third century BC, the level of V and stress had been falling for perhaps three hundred years Although C may rise for a time while V and stress is falling, as happened in both Europe and Japan and for a similar period of time, the fall of V and stress must eventually undermine C and cause both to decline. In Rome after 200 BC the decline was further driven by the enormous wealth brought by conquests following the Punic Wars of the mid to late third century.[236] The Roman writer Sallust had no doubts on the enervating effects of wealth, as he observed:

> When Carthage, the rival of Rome's sway, had perished root and branch, and all seas and lands were open, then Fortune began to grow cruel and to bring confusion into all our affairs … Hence the lust for money first, then for power, grew upon them; these were, I may say, the root of all evils. For avarice destroyed honour, integrity, and all other noble qualities; taught in their place insolence, cruelty, to neglect the gods, to set a price on everything …

> As soon as riches came to be held in honour, when glory, dominion and power followed in their train, virtue began to lose its lustre, poverty to be considered a disgrace, blamelessness to be termed malevolence. Therefore as the result of riches, luxury and greed, united with insolence, took possession of our young manhood. They pillaged, squandered; set little value on their own, coveted the goods of others; they disregarded modesty, chastity, everything human and divine; in short, they were utterly thoughtless and reckless.[237]

In the early second century BC, as mentioned earlier, Cato's sternness and integrity were already considered old-fashioned. Powerful aristocrats came to be known more for generosity and charming manners than for austerity and reserve, suggesting a switch to more personal, lower C attitudes.[238] There is further evidence of falling C in the second century. The birth rate fell rapidly despite efforts to maintain it. Peasant farmers flocked to the city, to be replaced by vast estates tilled by slaves. Even when the urban poor were given lands, as by the reforms of Gaius Gracchus later in the century, they sold them as soon as possible and drifted back to the city. As C declined, Romans no longer had as much taste for the drudgery of farming. Declining C and V also meant they had less taste for warfare, the consequence of which was a shift towards a professional army.[239]

Restraints on sexual behavior began to loosen. Divorce, uncommon in the early Republic, became acceptable by the first century BC. By the following century among the propertied classes it was almost routine.[240]

One obvious consequence of this change in temperament and economic behavior was a growing gap between rich and poor.[241] The elite were able to take advantage of new opportunities by buying up land, lending money and plundering the newly conquered provinces. But the great mass of Romans, less suited to productive work than their ancestors, became increasingly dependent on government-subsidized hand outs of food.[242]

There were also signs of a continued fall in V and stress in the second century BC. Women gained more power and were able to manage their own financial affairs and control property.[243] The law codes were relaxed and army discipline became more lenient.[244] The late Republic also saw a reduced harshness and a growing sentimentality towards children which would have contributed to flexibility of thought, since the punishment of older children creates more rigid attitudes through child V. Thus, business boomed even as the work ethic declined, and Latin literature also experienced a golden age in the first century BC. Writers and poets of the period such as Livy, Virgil, Cicero, Catullus and Caesar have been read and revered for more than two thousand years.

The decline of V and stress, which is associated with hierarchy, also gave birth to a more democratic sentiment. At the end of the second century BC the popular party, led by Tiberius and Gaius Gracchus, gained notable victories over the Senate, especially with regard to land redistribution. Although the Gracchus brothers were both murdered, when Julius Caesar gained power in the mid first century BC it was as leader of the popular party against the Senatorial aristocrats. While many of the developments associated with the early stages of declining V (especially child V)—free thought, creativity, democracy, freedom for women—can be regarded as positive, and there are obvious parallels to the modern West, there were more ominous signs.

By the beginning of the first century BC the loyalty of soldiers was turning from state to commander, which opened the way for successful generals such as Marius and Sulla to seize control.[245] Sulla's march on Rome in 88 BC was a pivotal moment in the fall of the Republic. It was an unprecedented action, and such a breach of republican principles that all but one of his commanders refused to follow him. In effect, their loyalties were still impersonal enough to give priority to the rule of law. But, and this is what mattered, the ordinary soldiers were now more loyal to Sulla than to the Republic. As a result, after a further incursion in 82 BC he was able to dominate the city, killing thousands of his political opponents. Between these two events his rival Marius had also dominated the city with soldiers personally loyal to him. The *timing* of these civil wars reflects the turmoil of a G-90 period around 90 BC, and the resulting localism would also have been the reason for the Social War of 91–88 BC, in which Rome's Italian allies rose against it.

But the key problem facing the Republic was the continuing fall in C. Laws and institutions are impersonal and will only be strong when people's loyalties are also impersonal. When C falls below a certain level, republican forms of government can no longer be maintained. The plotters who killed Julius Caesar in 44 BC thought they were restoring the Republic, but killing one man did nothing to change popular attitudes. When loyalties become sufficiently personal, the only question is which man is able to attract a greater following and thereby take control (see Fig. 12.1).

The role of personal loyalties in the quest for power can be seen in the life of Mark Antony, a fairly typical politician of the first century BC. Plutarch's description of him is in marked contrast to his observation of Cato:

> Antony grew up a very beautiful youth but, by the worst of misfortunes, became friends with Curio, a man abandoned to his pleasures. To make Antony dependent on him, Curio plunged him into a life of drinking and dissipation, resulting in so much extravagance that at an early age he went two hundred and fifty talents into debt.

> Antony's generous ways, his open and lavish gifts and favours to his friends and fellow-soldiers, did a great deal for him in his first advance to power. And after he became great, the same generosity maintained his fortunes, when a thousand follies were hastening their overthrow ...

> Antony did not take long to earn the love of his soldiers, joining them in their exercises and for the most part living amongst them, and making them presents to the utmost of his abilities. But with all others he was unpopular enough. He was too lazy to pay attention to the complaints of

injured parties. He listened impatiently to petitions, and he had an ill name for familiarity with other people's wives.[246]

Fig. 12.1. Rise and fall of the Roman Republic plotted against changes in stress and C. The Republic fills a segment in Rome's civilization cycle. It was established when C rose past a certain point, and fell when it declined. Republican and democratic governments require high C and especially infant C.

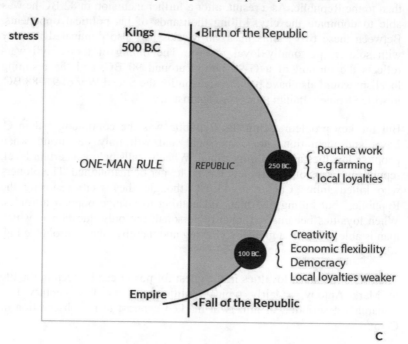

In this one man can be seen all the signs of lower C—extravagance, drinking, laziness, and sexual indulgence—but also a personal charm that endeared him to his soldiers and came close to winning him control of the Roman world. A greater contrast to Cato can scarcely be imagined.

The Roman Empire

Strictly speaking, in terms of territorial acquisition and international relations, Rome had been an empire since the second century BC when Greece and Carthage were brought under Roman rule. But historians date the Roman Empire from the new constitutional order established by

Augustus Caesar in 27 BC. He retained the form of Senatorial dominance, but took absolute power for himself.

The Empire brought political stability after decades of turmoil, aided by a new G year around 1 AD. But the new regime did nothing to halt the decline in C. Augustus was vividly aware of an apparent decay in the Roman character, and expressed it in this speech to Roman aristocrats in 9 AD:

> Mine has been an astonishing experience: for though I am always doing everything to promote an increase of population among you and am now about to rebuke you, I grieve to see that there are a great many of you [who are childless]. We do not spare murderers, you know ... yet, if one were to name over all the worst crimes, the others are naught in comparison with this one you are committing. Whether you consider them crime for crime or even set all of them together over against this single crime of yours. For you are committing murder in not begetting in the first place those who ought to be your descendants. You are committing sacrilege in putting an end to the names and honours of your ancestors; and you are guilty of impiety in that you are abolishing your families ... overthrowing their rites and their temples. Moreover, you are destroying the State by disobeying its laws, and you are betraying your country by rendering her barren and childless. Nay more, you are laying her even with the dust by making her destitute of future inhabitants.[247]

Augustus enacted legislation to try and turn back the clock, but was no more successful than Brutus and Cassius had been in their attempt to restore the old Republic through the assassination of Julius Caesar. Comments by Tacitus (55-120 AD) on the early Empire showed how the more old-fashioned Roman citizens regarded the new age. It is a familiar refrain:

> Promiscuity and degradation thrived. Roman morals had long become impure, but never was there so favourable an environment for debauchery as among this filthy crowd. Even in good surroundings people find it hard to behave well. Here every form of immorality competed for attention, and no chastity, modesty or vestige of decency could survive.[248]

Parents in the early Empire were less likely to punish their children, but also showed less interest in them and exerted less control. The children of the elite were often left in the care of slaves.[249] The historian Tacitus lamented the effect this had on the family:

> In the good old days, every man's son, born in wedlock, was brought up not in the chamber of some hireling nurse, but in his mother's lap, and at

her knee. And that mother could have no higher praise than that she managed the house and gave herself to her children. Again, some elderly relative would be selected in order that to her, as a person who had been tried and not found wanting, might be entrusted the care of all the youthful scions of the same house. In the presence of such a one no base word could be uttered without grave offence, and no wrong deed done ...

Nowadays, on the other hand, our children are handed over at their birth to some silly little Greek serving-maid ... The parents themselves make no effort to train their little ones in goodness and self-control; they grow up in an atmosphere of laxity and pertness, in which they come gradually to lose all sense of shame, and all respect both for themselves and for other people.[250]

The decline of Roman character was such that if the political and economic power of the Roman Empire had depended only on Italians it would have been in serious trouble by the first century AD. One indication of this is that Italian trade was already in decline.[251] But the genius of Rome was its ability to assimilate provincials. As declining C made Italians less capable, political power shifted to men from the provinces, who had been more recently acculturated and thus less affected by falling C. The great writers of the late first century AD, unlike those of the first century BC, were no longer Italians. Seneca and Lucan were from Spain, and Tacitus from Gaul. The economic and administrative life of the Empire was also increasingly in provincial hands. The efficient bureaucracy of the Emperor Claudius (41–54 AD), for example, was run by ex-slaves.

The same pattern can be seen in the emperors, though with a significant delay. Until Nero in 68 BC they came from ancient Roman families. Vespasian and his sons (69–96 AD) were from provincial Italy. Trajan and his three successors (98–180 AD) were from Spain and Gaul, and the successful emperors of the late third and early fourth centuries from mountainous Illyria.

The army showed the same pattern. Legions were increasingly formed from the peoples of conquered territories. By the fifth century the most effective soldiers were barbarians from beyond the frontier.

The Roman culture of the Empire, like that of the West in the twentieth and twenty-first centuries, was one of declining C. This made it attractive to Provincials, not only because of its prestige but also because it minimized the role of C-promoters which require people to act in ways that may be at odds with their temperaments. People accepting it would quickly lose V and thus stress, making them more flexible in thought. And because they initially maintained a higher level of C they became effective administrators, writers and even Emperors. But after a generation or two they lost C and sank into insignificance, allowing other groups from poorer areas to become prominent in turn. The rapid decay of character is especially evident in Imperial families, which were exposed to the extremes of wealth. The austere and hard-working Augustus was eventually succeeded by debauched libertines such as Caligula and Nero, the hard-working and self-disciplined Marcus Aurelius by the arbitrary and capricious Commodus (180–192 AD). None of these successors had much taste for the routine of administration.

Meanwhile, though on a slower timescale, C was declining throughout the entire Empire. We have seen that loyalties were already becoming more personal in the late Republic, allowing leaders like Marius, Sulla and Caesar to seize power.[252] Then the Republic died and the Empire was born, but for a time the old republican institutions—the Senate, the local bureaucracies—maintained some force. Magistrates were elected in Rome until the time of Tiberius (14–37 AD) and in most other cities until the second century.[253] But by the early third century AD the Roman state was an undisguised absolutism.[254] When Septimus Severus died in 211 AD his last words to his sons were: "Stick together, pay the soldiers, ignore everyone else."[255]

As loyalties became more personal still, even the emperor was too distant and impersonal to serve as a focus so the power of local leaders increased. One result was the growth of large estates that were in many ways immune from government control. Local magnates defied the law by evading taxes, keeping prisons, seizing debtors and arming retainers.[256] In some cases even provincial governors were helpless to stop them. The remaining authority of the emperor came to rely more and more on his actual presence and less on the laws and institutions of previous times. A distant emperor in Rome could no longer demand respect. Now he had to be physically present to exert his authority. In the fourth century the Empire was effectively split, with each half having an Augustus or senior emperor, and a Caesar or junior emperor.[257]

Lemming Cycles

But while the decline of C was continuous, the decline of authority was less so. The Empire came close to collapse in the early third century AD with a rash of civil wars and invasions. There was a marked recovery in the early fourth century under Emperor Constantine, who consolidated and Christianized the Empire.

This crisis and recovery is yet another lemming cycle pattern, with a G-90 period in 250 AD and a new G year around 340. The great plague of 165–80 AD in the reign of Marcus Aurelius fits quite well a G-150 year of 190 AD, which is the stage of the lemming cycle when resistance to disease is lowest. It was probably the worst epidemic to hit the Empire until the reign of Justinian in the mid-sixth century, its effects amplified by the fall in C which would have reduced resistance to disease (see chapter two). This means that G periods for the late Republic and Empire can be placed at 280 BC, 1 AD and 340 AD, very similar in length to the lemming cycles of Europe and Japan.

But note that while lemming cycles in Europe and Japan became shorter as C reached its peak, Roman lemming cycles became *longer* as C declined. And the next lemming cycle was to be longer still, with a G period after 440 years in the reign of Charlemagne (780 AD). Human lemming cycles are more variable than those of lemmings or muskrats, but they vary in an entirely consistent fashion.

The Long Decline

The G period of 340 AD provided only a short respite. By the early fifth century, loyalties had become so personal that the Western Empire collapsed. But this was less a single catastrophic event than a continuation of trends that had begun more than five centuries earlier.

The condition of the Empire had become even more precarious because of population decline, another indicator of falling C and V. The birth rate in Italy had been in freefall since the late Republic, and by the second century AD the population seems to have been falling in the Empire as a whole. Epitaphs show a large proportion of childless couples, partly because women seem to have been less willing to bear children.[258] In the early second century Pliny wrote that his was an "era in which the rewards of childlessness make many regard even one child as a burden."[259]

By the third and fourth centuries AD the fall in population was catastrophic. Barbarians were being settled in the frontier areas in the third century, and even in depopulated regions of Italy and the Balkans in the fourth. But despite this, the area of land under cultivation kept shrinking, especially in the west.[260] A shortage of labor is indicated by legislation tying tenants to the land, the willingness of landlords to employ runaways in spite of severe penalties, the extreme reluctance of landlords to release men for the army, and apparently modest rents.[261] The cities suffered even greater losses from flight to the country, and some disappeared entirely.

Falling C is also indicated by economic decline. Industrial activity in the ancient world is vividly evidenced by metal concentrations at the Greenland icecap (Fig. 12.2), which show that the biggest decline in metal production happened *before* the Empire collapsed.

Fig. 12.2. Deposits of lead and copper pollutants in the Greenland ice cap.[262] Economic decline, a consequence of falling C, started long before the Empire fell.

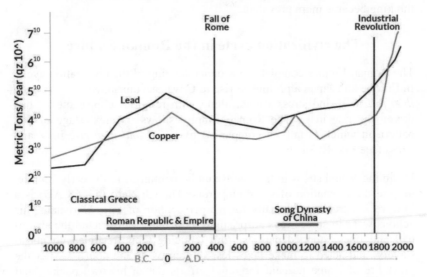

Trade also declined in the second, third and fourth centuries AD.[263] In the later Empire there was an emphasis on small-scale manufacturing for the local markets, and large estates were becoming more self-sufficient.[264] Taxes began to be levied in kind, even under a strong ruler such as Diocletian who made every effort to restore the currency.[265] It was the same problem as faced by Japanese rulers of the seventh century AD, who had tried to introduce coinage on Chinese models but failed. Currency simply cannot work in a low C society.

Diocletian and some fourth century rulers showed typically low C attitudes by restricting the market economy. Diocletian encouraged guilds and tried to control prices and the movement of labor.[266] All this is exactly what we would expect from a loss of the impersonal attitudes and flexibility associated with higher C.

Creativity followed the same pattern. We saw in chapter five that original thinking is especially characteristic of infant C, though inhibited by the more traditional thinking of child V. There was an initial burst of creativity in the last two centuries BC, following the initial decline of V, but after that it went into freefall as infant C declined. Thus, Latin literature went from the golden age of the late Republic to the silver age of the first century AD to the creative wasteland of the later Empire, in which there were few if any writers of note.[267] As infant C declined, superstitious thinking became more prevalent.[268]

The civilization cycle in the Roman Empire

The Roman Empire completes our understanding of the civilization cycle. In Europe and Japan a prolonged rise in C was accompanied by a *rise and then fall* in V and stress which, along with religious C-promoters, had driven the rise in C. The Roman Empire shows the next stage of this pattern in which C falls, accompanied by a *fall and then rise* in V and stress (see Fig. 12.3 below).

Declining V and stress in the late Republic continued in the early Empire, despite the reputation of some emperors. Though Nero (54–68 AD) is a byword for cruelty, especially for his persecution of early Christians, his actions were in some ways surprisingly humane. He restricted the size of bail and fines, stopped patrons from reducing their freedmen clients to slavery, and acted to make taxes less oppressive to the poor.[269] After the great fire of Rome he paid for relief efforts out of his own pocket, and spent days personally searching the ruins for survivors.[270] Unimportant men who insulted him were usually no more than banished.[271]

This trend toward lowered stress would become even more pronounced by the second century. Between 98 and 180 AD a series of humane and capable emperors did still more to alleviate harshness in the law.[272] Masters were no longer allowed to kill their slaves, castrate them or sell them to brothels without good cause.[273] The lives of citizens were guaranteed to a much greater extent, and even men guilty of conspiracy might not be executed.[274] And political stability, a key indication of low stress, was high. While six emperors were killed in the first century AD, none suffered this fate between 96 and 190 AD.

Fig. 12.3. The Roman civilization cycle. As C continues to fall, the decline in V and stress slows, and then V and stress begin to rise.

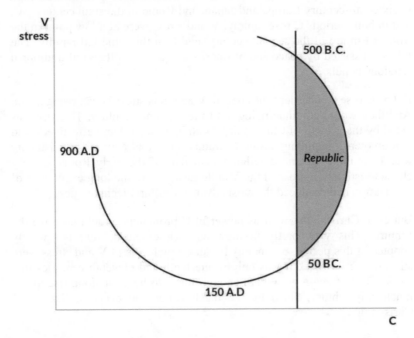

There were other indications of low V and stress in the early Empire. The freedom and power of upper class women seems to have been greatest in the first two centuries AD.[275] By the late first century some writers were opposing the beating of sons, and fathers no longer had power of life and death over their children.[276]

But from this low point, stress began to rise. From the late second century AD torture began to be used in some criminal trials. By the fourth century torture, flogging and such cruel deaths as burning were routine.[277] Deference and status differences became more marked and the ruler surrounded himself with pomp to emphasize his distance from other people, as opposed to the comparatively affable emperors of the second century.[278] The political system became more brutal and less stable. Commodus (180–92) and Caracalla (211–17) revived the reign of terror, with mass executions of political opponents, something not seen for almost a century. At the same time the position of emperor itself became more dangerous.[279] Commodus and most of his third century successors were killed in office.

The civilization cycle is symmetrical. V and stress were at an all-time high in sixteenth-century Europe and Japan, and Rome in the sixth century BC, and in both periods C rose rapidly. V and stress were at a low point in the Roman Empire of the second century, and C at this time fell rapidly. The fall was assisted by peace and prosperity and by the collapse of traditional religious beliefs.

But as C rose and harder times began, V and thus stress began rising again. And they would continue rising until the sixteenth century. This rise was aided by the growth of Christianity. With its powerful and effective V and C-promoters, including strong restraints on sexual activity, Christianity was a far more advanced religion than that of the early Romans. Very characteristically it arose in the Middle East, where the long experience of civilization had produced the most advanced cultural technologies.

But even Christianity, with its powerful C-promoters, could not save the Empire. This was, partly because its ascetic tenets were not widely adopted at this period, but mainly because low levels of V and stress were still driving down C. It was only in the tenth and eleventh centuries that the level of V again grew high enough for C to begin its long rise to the nineteenth century, aided by the much wider and deeper influence of Christianity.

The Fall of Rome

But before this happened there would be a long night of darkness, barbarism and strife. It is hard to comprehend the shock felt by the people of the ancient world at this catastrophe. Rome had endured for more than a thousand years, and for almost half this time had ruled much of the world. It brought unprecedented wealth and prosperity, magnificent cities, huge advances in technology, and a sophisticated literature and culture. Its language, the dialect of a small Italian city state, had become the native tongue of millions. Tourists and merchants could travel thousands of kilometers in relative comfort and safety.

And now it was gone, leaving utter misery in its wake. Populations collapsed to a fraction of their former levels, ridden with disease as rampaging armies marched back and forth across the land. Literacy all but vanished outside the monasteries, and much of the ancient learning was lost forever.

And yet, as we have seen, the fall of the Roman Empire was not a singular event that occurred in the early fifth century AD. Rather it was a long process that had begun more than six hundred years earlier. A high level of V had combined with powerful C-promoters, taken from the Etruscans and Greeks, to raise C to a peak in the third century BC. But V had been falling since around the sixth century, and by the third century it was no longer high enough to maintain C. Combined with the wealth brought in by foreign conquest, C began its long decline. The fall of Rome was deferred for many centuries by the Roman talent for assimilation, bringing in peoples with higher V and C as administrators, poets and soldiers until their own V and C was sapped in turn. Eventually, V reached a low point in the third century AD and began to rise, but this was too little too late. Even when combined with the powerful C-promoters in Christianity, C continued to fall until the Empire disintegrated.

Thus the seeds of Rome's fall were sown even *before* it reached the peak of its power and influence. And as will be seen in future chapters, the same can be said of the modern West.

The Eastern Empire

We have said that the Roman Empire "fell" in the fifth century AD, but this is not totally accurate. Certainly the Roman Empire in the West collapsed into chaos, but the East did not. The Eastern Empire enjoyed much greater political stability in the late fourth and early fifth centuries, with less of the civil conflict and usurpation that weakened the West at the time of the barbarian invasions.[280] Population even seems to have grown there in the late fourth, fifth and early sixth centuries.[281] And while Greece and the Balkans were overrun for several centuries, the provinces from Anatolia (modern Turkey) to Egypt were not. The Byzantine Empire was to continue as a major force in world politics for another thousand years.[282]

If Rome collapsed because of a long-term fall in C, then it seems that C must not have declined to nearly the same extent in the East. The reason for this is something called the S factor, which is the subject of the next chapter.

CHAPTER THIRTEEN

THE STABILITY FACTOR

Freedom is the sure possession of those alone who have the courage to defend it.
—Pericles

The puzzle of stability

The last chapter examined the fall of Rome, showing how a collapse in V and C undermined that once great civilization. This chapter will look at an equally important question—why long-established civilizations become more stable with time. The Eastern Roman Empire, which included areas with a long history of civilization, did not fall in the fifth century AD. It survived for another thousand years until taken over by strong successor states. The next chapter will show that ancient China and India went through their own process of collapse, though in a form far less severe than that of Rome, and both also gave rise to more stable civilizations.

The answer can be found in infant C, that aspect of temperament arising from parental control in early childhood and especially infancy. The Greek and Roman civilizations had characteristics of high infant C, as did those of ancient India and China. The peoples of the Eastern Roman Empire, and of India and China in later centuries, did not.

The reason to be proposed is that the process of civilization collapse causes people to go through a genetic change, which makes them more indulgent of infants and less so of older children. This means that infant C and thus C do not rise so high, V is less easily undermined (since indulgence of infants is a key driver of V), and the people are more conservative and less inclined to become wealthy. The genetic change happens because, in the process of collapse, people indulged as infants are more resistant to falling C and thus tend to have more children of their own. The resulting societies are thus less likely to experience a decline in V and C.

Thus, in sharp contrast to earlier chapters which focused on epigenetic changes, this chapter introduces a *genetic* change which has profound *epigenetic* consequences.

Survival of the Eastern Empire

The late Roman Republic and Empire experienced a prolonged fall in C which caused loyalties to become increasingly personal and then local, until by the fourth century it was no longer possible to hold the Empire together. Despite many attempts to restore unity, it was repeatedly divided into eastern and western halves. The final division, with separate emperors and governments, came in the mid-fourth century. From this point the Western Empire collapsed while the East did not.

One could attempt to explain the very different fates of the two empires in terms of economics and urbanization. The Eastern Empire was fully as wealthy as the West and had huge commercial cities such as Constantinople and Antioch. But unlike the West and especially Italy, there were few vast slave estates and peasant farmers continued to form the basis of agriculture. In biohistorical terms it is significant that the C-promoting practices of Christianity were more keenly adopted in the Eastern Empire. But as we know, Christianity was not at this stage a sufficient brake on the decline of C and V, either in the east or the west. Why, then, did the Eastern Empire survive?

Part of the answer is that C was never as high in the Eastern Empire. People with the highest levels of C are capable of impersonal loyalties which support republican institutions of the kind that arose in Rome and in Greece up to the fourth century BC. But the peoples of Anatolia, Syria and Egypt had no such institutions. Prior to their absorption into the Roman Empire they had been ruled by kings.

This is in itself significant, given that C is a product of C-promoters and the strongest C-promoters were from the Middle East, an area with a long history of civilization. Christianity originated there and so did the religions with which it competed, including the Egyptian, Isis cult and Mithraism from Persia. The same pattern can be found in Japanese history, where the cultural technologies underpinning C and V arose from the older centres of civilization—Buddhism from India and Confucianism from China.

This should not be surprising. C-promoters are driven by the needs of civilization, and the oldest centres of civilization had time to develop more powerful and effective forms. Societies with the strongest C-promoters could form larger states with harder-working and more productive farmers. Over thousands of years of competitive struggle it was inevitable that more powerful C-promoters and V-promoters should arise. But why then did the highest levels of C appear in civilizations such as ancient Greece and the Roman Republic, which were far distant from the ancient heartlands? The answer is that their levels of C could rise much higher because they also had very high infant C.

Infant C and the rise of the new civilizations

Infant C is a trait found in adults who experienced parental control in early childhood and especially infancy, as discussed in chapter 5. People with high infant C tend to be flexible and creative, with strong impersonal loyalties and an aptitude for engineering and machinery. Their temperament suits them for living in nation states. They also have an attachment to hereditary status, even when not associated with actual power (such as figurehead monarchs and dispossessed aristocrats).

As noted in chapter five, Japanese and Europeans tend to control infants but the peoples from the oldest centres of civilization, such as India, China and the Middle East, do not. Hence the rise of Rome and Greece with their impersonal systems of government, and advanced technology. Reverence for hereditary status can also be seen in the institution of the Roman Senate, where certain patrician families held power and prestige for five-hundred years.

They also benefited from exceptional military discipline. The key weapon of Greek hoplites was the shield, fastened to the left forearm in what was known as the 'Argive grip' which made it more effective in battle. But it covered only the left side and required fighting in strict formation, with each man protected by the man on his right.[283] Greek and Roman soldiers were no braver then their opponents and their arms could easily have been duplicated, but the ability to fight in this way required a certain kind of temperament. All these features can be explained in terms of a high level of infant C, which allowed the Greeks to conquer the Persian Empire and the equally disciplined Romans to take control of the Mediterranean.

The same pattern played out later in Europe. The Industrial Revolution was the direct result of very high C and especially infant C. And it is significant that the areas of Europe first to industrialize included lands never colonized by Rome such as Germany, the Netherlands and

Scandinavia. They also included provinces of the Empire where the spread of Germanic peoples was greatest, including Britain and north eastern France. Within Italy, industrialization was strongest in the north, especially the area settled by and named after a Germanic tribe, the Lombards. Areas where the native populations had not been displaced and where the population descended from the peoples of the Roman Empire, such as southern Italy and Sicily, remained economically backward.

In other words, based on the evidence of social and economic practices, a high level of infant C appears to be associated with civilized peoples who *have not previously experienced the collapse of a civilization*. By contrast, peoples who have undergone a civilizational collapse have lower levels of infant C. From this it may be inferred that the peoples in the oldest centres of civilization who were incorporated into the Roman Empire did not tend to control their infants.

Infant C, it seems, makes civilizations brilliant but also *unstable*. The reasons are clear. Infant C is associated with independent thought and rejection of traditional ideas. This can produce dramatic results in terms of technological development and wealth creation. But it also makes it more likely that people will abandon C and V-promoters. This is exactly what happened to the Manus people after European contact, when they abandoned their strict code of sexual morality. Infant C also makes societies wealthy, which tends to undermine their V and C and thus make them vulnerable to decline. In addition, control of infant tends to undermine V directly since it conflicts with infant indulgence, a factor that explains the decline of V in the civilization cycle (see chapter seven)

By Roman times it may be proposed that the peoples of the East, from modern Turkey down to Egypt, had developed a form of childrearing behavior that produced lower infant C and thus a much more stable society. The same thing would happen in China and India, as indicated in the next chapter. This behavior has survived and been observed in the recent past. Chinese and Egyptian villagers, though thousands of kilometers apart, show strikingly similar family patterns, including a tendency to be indulgent of infants and young children (see chapter five). In their traditional society they had high child V but low infant C and thus only moderate C, since the highest level of C requires infant C to also be high. It is this pattern of child rearing that makes societies more resistant to decline.

Introducing S—the Stability factor

Why would this be the case? Why would people who survive the collapse of a civilization indulge infants more than those who have not? And why would this behavior become so entrenched? The answer seems to be that it is rooted in the genes. In other words, we are looking at a genetic change resulting from the collapse of a civilization.

At this point it is necessary to ask how people in the oldest areas of civilization avoid controlling their infants or young children, since biohistory has to this point explained such control as the expression of C, a complex of attitudes and behaviors that naturally leads people to control and discipline children.

The patterns of behavior that support C and V, including patriarchy and restrictions on sexual behavior, can travel from culture to culture very fast. As indicated in the last chapter, the Romans presumably adopted such practices from the Etruscans who settled to the north of them and from Greek settlers in the south. Within a couple of centuries Roman culture had emerged with both patriarchy and female chastity at very high levels.

The spread of Christianity was even faster. In only three centuries it conquered the Roman world and then spread to the barbarians. The Vandals and Goths who invaded the Roman Empire in the fifth century were already Christians, though somewhat different in beliefs and practices than their counterparts in the Empire.

But the tendency to indulge infants while being more severe with older children—the practice which suppresses infant C, inhibits civilizational advance but promotes stability—does not spread in the same way. For example, nineteenth century Europeans retained their urge to control infants despite 1,500 years of exposure to a Middle-Eastern cultural technology—Christianity. They did not adopt that region's indulgence of young children. The same can be said of the Japanese, who adopted the C and V-promoters of Buddhism and Confucianism but not the infant indulgence of the Chinese and Indians.

The same applies in reverse. The Greeks were culturally dominant in the Levant for several centuries after Alexander's conquest, and many local peoples accepted their educational systems and even language. But there is not the slightest trace of early control behavior in the conquered peoples, such as the development of nationalism or the unusual creativity associated with infant C. In other words, the tendency to indulge infants does not seem to be cultural as such.

Instead, it appears to be a genetic feature that inclines people to indulge and protect babies and very young children, thus overcoming any inclination to control them (Fig. 13.1). It is probably similar to the aspect of baboon behavior that causes this otherwise highly aggressive species to protect and nurture infants, triggered at least partly by their distinct black colouring. Biohistory refers to this as the "Stability Factor" or "S," since its most obvious effect is to make societies more stable.

Fig. 13.1. Low S and high S

Low S cultures with high C tend to train children in both infancy and later childhood

Training 0-5   Training 6 and up

High S cultures always indulge children in infancy and early childhood, but may train them rigorously at later ages

Indulgence, no training 0-5   Training 6 and up

It is quite plausible that such a genetic trait could increase dramatically in a few hundred years, since other genes are known to have changed under the selection pressures of civilization. For example, the intelligence of Ashkenazi Jews appears to have risen significantly over the course of about five centuries due to the selection pressure caused by their mercantile role.[284] Less intelligent people either failed to have so many children, or perhaps ceased to be Jews. There is also evidence that farming has accelerated the pace of genetic change a hundredfold in areas such as disease resistance and the ability to digest lactose.[285]

The selection pressures driving higher S were at least as strong as these, since higher S makes it likely that people in a wealthy urban society will have children. This is because, lacking infant C, they tend to be poorer and more conservative. Thus it was that the higher S peasants of the Eastern Roman Empire stayed on their farms and bred children, while the lower S peasants of Italy flocked to the cities and (by and large) did not. The result was that the people of the East maintained enough C and V to support strong central government and military effectiveness. Those in the West did not.

Just as high-S societies tend to maintain a higher birth rate than low S societies, so do high-S people within a low-S society. In twentieth-century Italy the birth rate dropped earlier and faster in the industrial north, where more low-S Germans had settled after the fall of Rome than in the agricultural south. Though it must be remarked that once a people becomes wealthy enough, even high S is no defense. The Sicilian fertility rate is now down to 1.37 children per woman, well below replacement level though still higher than the 1.28 fertility rate of northern Italy.[286]

This analysis also makes it clear not only how but *when* the level of S begins to rise. It is not experience of civilization as such, since nineteenth-century Britons had been civilized for almost a thousand years and were clearly low S. This is what made the Industrial Revolution was possible. In fact, up to this time the selection process would have been, if anything, to lower S. The higher infant C made possible by low S is a major contributor to economic success, and we have seen that wealthier people tended to have more surviving children before the modern era.

But as soon as affluence sets in and C and V start to fall, the advantage shifts to higher S. A graph of how this might look is given in Fig. 13.2 below.

Why does this genetic change matter?

Biohistory is epigenetic rather than genetic. Patterns of control and punishment of children at different stages of life result in epigenetic changes which influence economic activity, creativity, attitudes towards religion and scholarship, forms of government, the size of political units, and much more. So why the focus on high versus low S? If all S does is determine how people treat their children, should we not simply focus on the actual behavior and the epigenetic changes that result from it? This was essentially the approach taken in chapter five, when infant C was considered one of a number of epigenetic variables.

Fig. 13.2. Change from a low-S to a high-S society following civilization collapse. During a civilization collapse, people with the genes for higher S are more likely to indulge infants. This gives them lower infant C and higher child V, so they retain traditional values and tend to have more selection. Thus, high-S genes spread in the population.

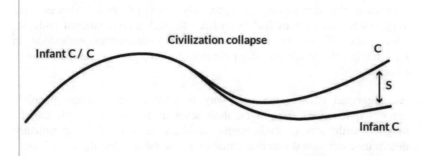

S matters because it *cannot* be changed epigenetically and thus follows a different pattern of evolution. It is not affected by culture and does not spread through religions and philosophies. This difference explains many of the key questions of history, such as why the Eastern Roman Empire did not collapse with the fall of Rome. It shows why industrial civilization developed in Europe and especially northern Europe, and not in the Middle East or India. The reasons are rooted in the epigenetic effects of treatment in infancy, but understanding S explains why that treatment happens.

In certain important ways, S acts contrary to other trends in human history. A popular view is that humanity is moving towards ever greater achievements in knowledge, power and economic activity, threatened only by ecological factors such as climate change or the exhaustion of key resources such as oil and tropical forests. But rising S puts a check on human progress by acting to curb, in the long term, both creativity and economic growth in order to make a society more stable biologically. This means that if and when Western civilization collapses, the successor civilization will be poorer and less brilliant, but more stable.

For all these reasons it is vital to understand what high and low S societies are like, and how one changes into the other.

Low-S societies

Europe is the classic low-S culture, which is what made possible the rise of infant C to unprecedented heights in the nineteenth century. This in turn made possible the Industrial Revolution, along with the extraordinary creativity and organization that enabled Europe to dominate the world. It now seems that Europeans, and especially those of Germanic descent, had very low S because they had never been through a civilizational collapse. Thus, when C increased as a result of C-promoters embedded in Christianity, the result was strict control of young children and a surge in infant C.

As suggested earlier, the best way to explain this is that people's emotional reactions depend on their level of S. Low-S people can be tougher with infants and young children because their genetically determined emotional reaction causes them to be less indulgent to infants than are people with higher S. It is also possible that they are genetically predisposed to be less strict with older children.

As indicated in chapter five, this was a Europe-wide pattern and one reaching back to at least the fifteenth century. The contrast between the childrearing behavior expected of European and American parents early in the twentieth century, and that of Chinese and Egyptian, parents, could hardly be more extreme. In China and Egypt, a child was barely considered able to understand right from wrong until five or six years old. In Europe, strict discipline started from the first year of life. Societies with high infant C can be readily identified by such features as republican institutions, stable nation states within the culture area, and unusual wealth and creativity.

Low S and respect for hereditary rank

It was also mentioned in chapter five that one distinguishing feature of low-S societies such as Europe and Japan is an unusual respect for hereditary rank, even when the holder of that rank lacks any real power.

Many European nations have constitutional monarchies, in which the monarch has no actual executive power. Great Britain is perhaps the best known but Belgium, Denmark, Liechtenstein, Luxembourg, Monaco, Holland, Norway, Spain and Sweden all have powerless but highly regarded monarchs. Significantly, these nations (apart from Spain and Monaco) are in areas of northern Europe that were overwhelmingly settled by Germanic peoples after the fall of Rome, and thus have low S since none experienced civilization collapse.

The supreme example of a low-S non-European society is Japan, which had an unbroken line of figurehead monarchs for well over a thousand years. The Japanese have been so attached to powerless rulers that the shoguns, who originally received their authority from the emperor, retained their positions in eras when they were reduced to the same powerless status. This was especially so in the Kamakura period (1185–1333 AD) when Hojo regents ruled in the name of the puppet shogun, who ruled in the name of the powerless emperor.

European and Japanese respect for hereditary rank has allowed people from noble families to retain power even when new forms of government have developed. The European and Japanese civilizations have been characterized by influential and long-lasting aristocratic families which managed to co-exist with powerful and centralized states. The British aristocracy still held considerable sway in the nineteenth century, including families that had been prominent in national affairs for five hundred years and more, still occupying the upper ranks of government long after the development of democracy. The roll of nineteenth-century British prime ministers is filled with hereditary earls, viscounts and dukes. Even into the twentieth century parliament was dominated by men from aristocratic families such as Winston Churchill, who lacked a title but came from the family of the Duke of Marlborough. Ancient Rome, an early low-S civilization, showed the same pattern. Julius Caesar, who put an end to the Republic, was a direct descendant of a family which had been influential at its founding a half-millennium earlier. So too was Brutus, his chief assassin.

The long-civilized areas of China, India and the Middle East were also ruled by hereditary monarchs for thousands of year, but these rulers were usually killed or at least dethroned when they lost power. Dynasties such as the Ming and Qing held power for centuries, but none were retained as figureheads. A Chinese ruler was revered only so long as he was believed to be powerful. When the power left him, so did the reverence. Monarchs who reign but do not rule are quite rare in historical terms, so it is all the more significant that we find them *only* in societies with low S—in other words, societies which control their very young children.

This respect for hereditary status, even in the absence of real power, will be an important indication of low S civilizations in the past. It is especially useful because it appears hundreds of years before the peak of infant C.

Empires versus Nation States

Higher S societies, found in regions with a long history of civilization, are typified by the Egyptian, and Chinese parents described in chapter five. Such peoples readily accept a place in cosmopolitan empires rather than forming tight-knit nation states. When empires collapse, the dynasties and political borders of high-S peoples change rapidly. They normally have higher V and stress, which means people tend to accept powerful, stabilizing authority. The lack of impersonal loyalties and strong legal codes gives local elites considerable power, but such power is far less likely to be inherited.

Westerners typically see their own wealthy and innovative nation states, with high infant C resulting from low S, as more advanced and successful. In fact, the next two chapters will indicate that low S is a feature of young civilizations, and one which most civilized societies have passed through and abandoned. High-S societies of the modern world were originally low S, with all the characteristics that go with it.

The rise of S—Sicily and Southern Italy

In Europe the highest levels of S are found in former Roman provinces most distant from the frontier, which experienced civilization collapse but where fewer barbarians had the chance to settle. One such is southern Italy and Sicily.

The original inhabitants had no history of civilization and so were presumably low S but also low C. They were displaced or absorbed by Greek colonists from the eighth century onwards, after which the area then became known as *Magna Graecia* or Great Greece. The Greek settlers, with their creativity and city states, clearly had low S. There were also Carthaginian settlers originally from Phoenicia (modern Lebanon) who presumably had higher S, but genetically were probably a small part of the mix.

S may well have been rising before the Roman conquest, since Greek birth rates were falling as early as the fourth century BC, with further selection pressures brought about by the long Roman decline. But unlike Lombardy and areas further north with substantial Gothic populations, relatively few Germans settled in the south. In fact, the Roman aristocracy held on well into the sixth century AD. Thus, the whole subsequent history of this area shows the effects of higher S with lower infant C, in sharp contrast to what happened further north.

One of the features of such a society is acceptance of authority in general, with less of the loyalty to people of similar ethnicity and language which entrenches local regimes. Thus, in place of the city states of northern Italy or the nation states beyond, the south and especially Sicily was ruled by a succession of alien powers: Byzantines, Normans, Arabs, French, Germans, Aragonese and Austrians. The strength of the Mafia, perhaps Sicily's most famous export, derives from a culture which values personal ties above all, and sees government as an alien entity to which no loyalty is due—all characteristics of low infant C.

The south of Italy has remained poorer, more conservative and religious, and until recently had a higher birth rate than the north, further indications of lower infant C. In recent times the birth rate has fallen below replacement levels, as in the rest of Italy, which raises the crucial point that high S does not stop C and V from collapsing. It simply makes the process slower. Other things being equal, high S people are more likely to hold on to traditions and less likely to become prosperous. But given enough prosperity, which in biological terms simply means not going hungry, both C and V must fall.

The same general distinction between high and low S can be made between southern and northern Europe. The lowest S and highest infant C can be found in the northerly and Germanic-speaking areas. The distinction between Catholics and Protestants runs roughly along the same lines, the obvious exceptions being Austria and south Germany which remained Catholic after the Reformation.

An idea of how infant C and C may have changed in the Eastern Empire, the longer-civilized areas of the Western Empire, and the northern lands is given in Figs. 13.3 to 13.5 below.

Fig. 13.3. C and infant C in Egypt and Syria. Because infant C was already low, given long experience of civilization, the decline of C and child V in the Roman period was only slight and temporary, staying high enough to support strong empires.

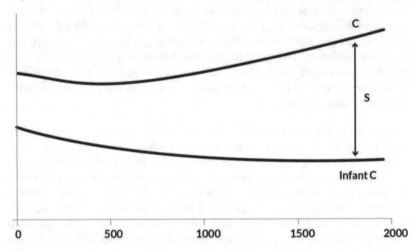

Fig. 13.4. C and infant C in southern Italy and Sicily. Because infant C was still relatively high in Roman times, the collapse of C and child V was more extreme, falling too low to support strong empires. The experience of collapse drove selection pressures for higher S so that infant C remained low, leading to a more stable but less creative society.

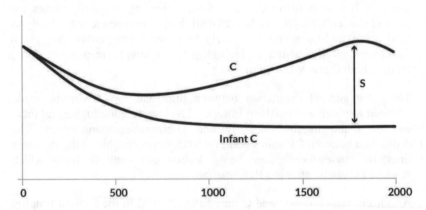

Fig. 13.5. C and infant C in northern Europe. The Germanic peoples did not experience a fall in C because they were not part of the Roman Empire, so S remained low. Thus when C rose under the influence of Christianity, infant C rose along with it. It was this ultra-high infant C which made industrialization and the rise of Europe to world power possible.

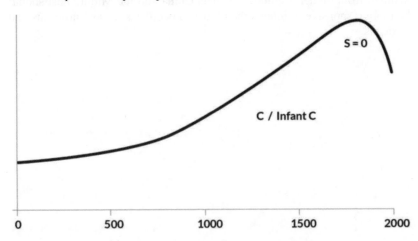

Western civilization has been enormously successful over the past two centuries, a success stemming directly from the early control of infants which increases infant C. This explains a number of the peculiarities of Western civilization—its creativity and technological brilliance, its democratic nation states, and its strong respect for hereditary rank. This success was made possible by powerful C- and V-promoting cultures imported from the earliest centres of civilization in India, China and the Middle East. However, early control of children—and the wealth stemming from it—also makes such civilizations unstable.

By contrast, the oldest centres of civilization have gone through what appears to be a genetic change that makes people more indulgent of infants and thus less likely to control them. The result is very low infant C, moderate C and high child V, producing societies less brilliant but more stable, and tending to form cosmopolitan empires rather than nation states.

The next two chapters will show that the earliest civilizations of India, China and the Middle East were originally low S, similar to those of Northern Europe and Japan. They will follow the processes by which the experience of collapse transformed them into the more stable high-S societies of the present day.

Testing

Measures such as pupil dilation could be used to assess the interest of people, and especially women, in babies and children of different ages. Women from higher S cultures such as China and the Middle East should be more responsive to infants than lower S peoples such as Europeans and Africans.

CHAPTER FOURTEEN

CHINA AND INDIA

Three things cannot be long hidden: the sun, the moon, and the truth.
—Buddha

In the previous chapter it was suggested that when civilizations collapse, people with a genetic predisposition to indulge infants (high S) are better at maintaining traditional values, because they tend to have lower infant C and higher child V. This means that they tend to have more children and high-S genes spread. The result is that when a new civilization rises it is more conservative and with less of the local and hereditary loyalties associated with high infant C. It thus tends to form cosmopolitan Empires rather than city states. This concept was used to explain the greater stability of the Eastern Roman Empire, which was already higher S and so survived the collapse of the Western Empire. It is also consistent with the relative poverty of areas such as Sicily, which survived the collapse of Rome without a mass incursion of low-S barbarians.

This chapter will trace the same process in ancient India and China, tracing their history from ancient low-S civilizations, through a prolonged collapse, to the more conservative and stable societies of recent centuries. It will also suggest that classical Greece, the epitome of high infant C in the ancient world, has gone through a less complete version of the same process.

An enlightening journey

A keen tourist born around 500 BC, with sufficient wealth and a taste for adventure, could have made a very significant journey. Starting in China as a youth, he might have met a would-be administrator who was moving from place to place among the states into which China was then divided, seeking a post worthy of his talents. Though unsuccessful in his main ambition, this man gathered along the way a number of students to whom he imparted his ideas. Known to us as Confucius, he became the founder of a tradition which has shaped and dominated Chinese culture to this day.

Travelling to India, the tourist might have come across two other men. One, Mahavira, was setting down the canonical texts for what would become the religion of Jainism. The other, Siddhartha Gautama, the son of a local prince turned monk, was spending his life wandering around the states in which northeast India was divided, teaching ideas that would go on to dominate Asia for thousands of years. He would become known to history as the Buddha.

Finally, moving to Athens as a middle-aged man, the tourist might sit at the feet of Socrates, another teacher who would have a profound influence on future generations.

It is a remarkable fact that these extraordinary figures all lived and taught within the lifetime of a single individual. We are accustomed to seeing the present day as a time of innovation and change, compared to the more static and unchanging past. For the West this has some validity, but for the oldest centres of civilization in India, China and the Middle East there is a strong argument that their most creative time was in the distant past. For India and China, that time was in and around the fifth century BC.

The oldest centres of civilization are characterized by their abilities to resist decline and collapse, suggesting high levels of S, the stability factor. This genetic trait causes people to be more indulgent to infants and younger children than to older ones, thus suppressing the creative infant C and promoting the conservative child V. S is a unique trait in that it rises in peoples whose ancestors have undergone a civilizational collapse. In this chapter we look more closely at the evidence for increasing S (and thus declining infant C) in the histories of China, India and Ancient Greece. We will see how those civilizations went from the radically inventive periods that produced Confucius, Buddha and Socrates, and evolved into more stable but less innovative societies. We will also uncover other reasons why S might rise in a post-collapse culture.

China

Present-day China is a high-S culture where infants are treated indulgently, as discussed in chapter five. A characteristic of such societies is that people are less resistant to rulers unlike themselves in language, culture and religion. This means that an empire which arises in a region of culturally related high-S societies can expand to fill the entire region, absorbing and ruling all the individual cultures despite local variations in language, values and so on, as in the Middle East from the late centuries

BC onward, and in India throughout most of the past several millennia. Similarly, the conquest of China by Mongols and Manchus produced two of the last three Chinese imperial dynasties. High-S societies like China also have weaker loyalty to hereditary leaders, meaning that although a hereditary system may be in place, its members will not be retained as figureheads if they lose their effective power.

This is in contrast to low-S societies such as those of Europe, which strongly resist rulers from outside their culture area (defined as a collection of societies with shared cultural traits but local variations, such as Europe). They will even resist rulers with relatively minor differences from within the culture area, leading to a proliferation of nation states and even city states. They are also more likely to retain figurehead monarchs.

The later history of China is that of a stable, high-S civilization, but it was not always that way. We know that high S is acquired during the process of a civilization's collapse, so we should expect the civilized Chinese of the earlier centuries BC to be low S and to have developed higher S in the centuries that followed. We also see civilization cycles playing out in Chinese history, including a collapse of C and V from the second century BC and a steady rise from the sixth century AD onward.

Low S—the Zhou dynasty

The earliest Chinese civilization occurred under the Shang dynasty, which ruled towards the end of the second millennium BC. [287] Around 1046 BC it was overrun by invaders from the Wei river area in the northwest. The invaders settled and established the Zhou dynasty, which controlled what is now northern China. For a few centuries they prospered, until in 770 BC their capital was sacked and the centre of Zhou power relocated eastwards.

From this date, the power of Zhou emperors declined. Even before 770 local rulers probably had a great deal of independence; and after this they became effectively independent. But it is a clear sign of low S that the powerless Zhou emperors were revered as overlords for more than half a millennium.

For example, after one Zhou ruler had been ejected from the capital by an army from the southern state of Chu in the mid-seventh century, he was restored in 635 by Chonger, ruler of the northern state of Jin. The grateful Zhou emperor awarded Chonger the title of *ba* (hegemon). The ceremony in which he accepted this honor, after refusing it twice as a sign of humility, shows the extreme reverence expressed by this powerful lord for his utterly powerless sovereign:

> Chonger ventures to bow twice, touching his head to the ground, and respectfully accepts and publishes abroad these illustrious and enlightened and excellent commands of the (Zhou) Son of Heaven.[288]

The emperor may have ceased to rule—except in his own small principality—but he continued to reign until well into the third century BC. This pattern, never seen again in China, is typical of the reverence for hereditary rulers associated with higher infant C, made possible by low S.

A significant characteristic of low-S societies is that hereditary aristocrats tend to play an unusually prominent role when C is moderate and loyalties still personal, compared to the more fluid and fast-changing elites in high-S societies. This was certainly the case in China where until the sixth century BC the great majority of prominent men were from great or royal families. There was a particular feeling that noble lines, which were responsible for ancestral sacrifices, should not be extinguished even when they lost power.

The period between the eighth and third centuries BC was a time of endemic warfare, but although larger states tended to absorb smaller ones, most retained their identity for hundreds of years. National loyalties were so strong that they could even survive more than a century of foreign conquest.

During the seventh century there are signs of rising C and especially infant C in China. States became increasingly centralized, with elaborate and often draconian law codes. This made possible much larger armies, which contributed to the growing ferocity of warfare. Due to their acceptance of impersonal authority, people with high infant C are relatively easy to organize and discipline by ethnically similar rulers, as in the Roman Republic or the modern West. So it was in China.

This went hand in hand with economic development. Until the fifth or sixth centuries BC, estates tended to be self-sufficient and trade was slight. Wealth was measured in livestock and salaries were paid in grain. But by the late fourth century farmers were buying (rather than making) their cloth, cooking pots and implements. Trade and transport conditions improved and coins were coming into use. Population grew dramatically and Chinese settlement expanded.

This was also a time of innovations in philosophy without parallel in Chinese history. Confucianism was established, along with the less savoury school of Legalism which sought to justify the harsh policies of the new centralizing states. In addition, popular beliefs were organized into what became known as Taoism, a quietist philosophy advocating retreat from the world. By the fourth and third centuries BC, change was no longer seen as a thing to be feared. Wherever they might be useful, new practices were adopted gladly without regard for past traditions.

The evidence is not detailed enough to pinpoint the peak of C, though it probably occurred around the fourth century BC since the exceptional openness of the fourth and third centuries is characteristic of the early stages of declining C. There are also indications of high V and stress at this time. Judging by political instability the high point of stress might possibly be placed in the fifth century when the rulers of Jin, Qi and possibly Sung were deposed by leading families within their state. It must be remarked, however, that rulers in the fourth and third centuries BC imposed exceptionally harsh punishments.

The Decline of C

But by the third century BC, Chinese C was in decline. A clear sign was the weakening of local loyalties, which made unity possible under the Qin and Han dynasties after 221 BC. In the following century Chinese power and wealth were at a peak. C was still high enough to support imperial authority, and unity made it strong enough to gain control over outlying lands. But as C continued to fall, the strong centralized state began to weaken. The tax-free estates of great families expanded rapidly, leading to a crisis in government finance after the mid-first century BC. The problem was not helped by a series of weak emperors, but the underlying cause was social rather than political.

There is also evidence of low V in this time, with mass armies composed of criminals suffering disastrous losses against the armies of the northern pastoralists. The harsh punishments and penalties of earlier times were relaxed, indicating a decline in stress.

The end of the Chinese Empire

By the end of the first century BC the imperial government was losing its grip. There was a brief recovery in the first century AD under the Later

Han dynasty, the consequence of a G period around 80 AD, but the great families with their tax-free estates and private armies remained largely outside government control. As in Rome, economic activity receded as the Empire declined. Over the following centuries the great estates became more and more self-sufficient, trade declined, and coinage virtually went out of use in some areas. At the end of the second century AD the empire collapsed into several centuries of chaos and disorder. It was to be by far the longest Dark Age in Chinese history.

Although later periods in Chinese history were marked by outbreaks of strife, the chaos of the third to sixth centuries AD was of a particular kind, the result of a collapse in V and C similar to that which caused the fall of Rome. The collapse was so extreme because ancient China, like ancient Rome and the modern West, was a low-S society with high infant C. This made it brilliantly successful in its heyday but also unstable.

Rising S

Despite clear similarities during the immediate post-collapse period, the later history of China was to be very different to that of Europe after the fall of Rome. The reason for this is that the Chinese population remained in place. The Roman population suffered a catastrophic decline under the Empire partly because of slave estates and also because the sheer brilliance of Roman administration kept peace and prosperity going for four hundred years, allowing V and C to decline much further than in China. Also, the barbarians were farmers and relatively numerous, and much of Europe remained out of the Empire. Thus the new Europe that emerged from chaos was largely descended from barbarians, especially in the north. They had never experienced a civilization collapse and thus remained low S.

By contrast, Chinese population decline was far less serious. There were no massive slave estates, and no dispossessed peasants flocking to the cities. And while the population of north China declined, that of south China actually expanded during the centuries of disorder. Also, the nomads of the northern plains were pastoralists far less numerous than Chinese farmers.

The rulers of the post-collapse regimes may have employed nomad horsemen to fight for them, but it is significant that the rulers themselves were native Chinese. Thus the new China that emerged from chaos was overwhelmingly Chinese by descent.

S rises when a civilization collapses because the fall of C and V gives a competitive advantage to people with higher S. As discussed in the last chapter with reference to Sicily versus northern Italy, being poorer and more conservative they are less likely to suffer a dramatic decline in birth rate. With fewer barbarians to introduce low-S genes, the Chinese experienced a significant rise in S. This was already obvious by the fifth century AD when a new dynasty unified north China for several decades. No stable nation states were established and there was no revered but powerless "Son of Heaven." Instead, dynasties rose and fell with startling speed.

Eventually, the empire was reunited under the Sui (589–618) and Tang (618–907) dynasties, and from then on strong dynasties alternated with periods of chaos. Rather than favoring aristocrats, as in a low-S society, these dynasties began selecting bureaucrats by competitive examination rather than noble birth. The Chinese had clearly changed from a low-S to a high-S people.

The re-emergence of high C

After the dark ages that followed the collapse of the Han dynasty, C began to rise again. The reunification of the Empire was one clear sign. Trade and the use of money revived during the sixth and seventh centuries and grew rapidly from the eight to the thirteenth. The order and functioning of cities came to be based on trade, rather than administration as in the past. Political stability increased, with shorter periods of disorder.

There was a gradual strengthening of C- and V-promoting traditions, presumably the result of intense competition. In peaceful agrarian societies where most people live at the edge of subsistence, farmers and merchants who work hard tend to have more surviving children than those do not. Thus it is that attitudes to sex became stricter with time, closely linked with the subordination of women and the resurgence of Confucianism. Whilst women in the Tang dynasty in the seventh century AD were relatively free, by the northern Sung Dynasty (960–1126) women began to be secluded and separated from men. The binding of girls' feet was introduced, and women began to be portrayed in a more fragile manner. Clothing became more modest. This process accelerated with the triumph of Neo-Confucianism in the thirteenth century and under the Mongols (1239–1368). Meanwhile, the Mongol conquest and the occupation of north China by barbarian invaders in the previous century worked to raise the level of V, thus strengthening the cultural technology that helped raise child V and C. The Mongols would have had low S genes, but were not numerous enough to significantly lower S in the Chinese population.

Most of our evidence applies to the elite, but Confucian attitudes to sex seem to have been spreading to other classes by at least the Ming Dynasty (1368–1644). In the Qing period (1644–1911) pornographic literature was suppressed and sex, according to some accounts, came to be seen as more a burden than a joy. Ethnographic studies of traditional villages in the twentieth century reveal an intensely conservative attitude to sex.

A continuing rise in S

Along with the rise in C there are indications that S also continued to rise. In other words, although C rose, infant C stayed low. Even though S was considerably higher after the collapse and rebuilding of Chinese civilization, it was still lower than modern levels. The Tang period did not show the philosophical ferment of the Warring States period, but it was brilliant and creative in literature and the arts.

Later centuries, especially from the thirteenth onward, were both less creative and more rigid and orthodox in their thinking, with few philosophers of note emerging after the twelfth century.

Another way to track the rise of S is to follow the decline of the hereditary principle. It was enormously important in the Zhou dynasty, and still of moderate importance in the Tang dynasty, with many if not most of the leading figures, including the Sui and Tang founders, coming from the north-western aristocracy. By comparison, elites in later times tended to gain wealth initially through trade or landholding, giving them the resources to train sons for the imperial bureaucracy, and to use this as a road to greater wealth and power. Success was achieved not by family connections but by success in the imperial examinations. There was still a hereditary system, in the form of the imperial structure, but the emperor stayed in place only so long as he held actual power. On this basis we can see a steady fall in hereditary loyalties and thus a rise in S from the Warring States period, to the Han dynasties, to the Tang, to the fourteenth century and after.

Rising S is also evident in China's military situation. Under the Han dynasty, as in the Roman Empire, infant C (made possible by low S) eventually undermined V and led to military collapse, aided of course by the breakdown of the state as a result of falling C. China also became militarily weaker during the Song period. It never even controlled the area around Beijing and from the early twelfth century lost north China and then (a century and a half later) the south to the barbarians. But this military weakness was not the same as that of the collapsing Han dynasty. The Southern Song dynasty resisted the extraordinary might of the Mongol Empire for some time, and later regimes were still more successful.

The Ming, Qing and Communists did far better than the Song in controlling the non-Chinese areas to the north and west. This is an indication that V collapsed drastically under the Han, much less under the Song, and still less in later periods. All of this is consistent with a steady rise in S, limiting the level of infant C, which in turn allowed V to be maintained.

The question must then be asked as to *why* S continued to rise, given that China did not again experience the collapse in C and V that is presumed to have triggered its original increase.

The reasons for this may well be political. In a low-S society with strong feudal and then national loyalties,   such loyalties were no hindrance to success and survival. But once S had risen during the collapse and the balance shifted towards cosmopolitan empires, intense local loyalties could be dangerous and even suicidal. Given the bloody vengeance inflicted by emperors on rebellious subjects, it might be wiser and safer to submit to the strongest power around.

India

Our tourist's second stopping point is in north-eastern India, where another low-S civilization was flourishing by the fifth century BC.[289] It comprised 16 states covering the Ganges plain, north to the foothills of the Himalayas, and north-west into modern Punjab. Many of these states kept their identities for several hundred years, and some had republican forms of government.

This was a time of prosperity and growth. Trade expanded, cities grew, and money was beginning to circulate. It was also a time of innovation in thought, with a widespread reaction against traditional beliefs involving ritual and animal sacrifice. As mentioned earlier, the founders of both Buddhism and Jainism lived at this time. Others schools of thought were deterministic (believing that the future was predestined and nothing could change it), or totally materialistic (believing in no gods at all). Buddhism and Jainism, especially the latter with its asceticism and emphasis on moral integrity, were highly ethical religions. These features also indicate high infant C made possible by low S.

The collapse of ancient Indian civilization

It is also indicative of low S that this civilization proved unstable. At the end of the fourth century BC it coalesced into the Mauryan Empire which soon came to control most of the subcontinent. By the third century BC there were increasing signs of pacifism, a clear indication of falling V. Though the Mauryan Empire reached its peak under Ashoka Maurya, who reigned from around 269 to 232 BC, it disintegrated rapidly after his death. The last Mauryan ruler was assassinated in 185 BC, to be followed by nearly six centuries of disorder and nomad invasions. Again as in ancient China, this strongly suggests a civilization collapsing as the result of falling C and V.

Emerging from this extended period of anarchy, the India of the fourth century AD onwards was not one of nation states like those of the earlier Ganges civilization (1500–500 BC), but of strong empires alternating with periods of disorder. Compared to China it had less experience of empire and more of disorder, but the overall pattern was the same. Also, like China it was increasingly conservative.

All this indicates higher S, which rose for the same reason it had in China. India under the Mauryans was not radically depopulated, as was the Roman Empire, and the barbarians were relatively few. Thus when C and V collapsed during the Mauryan period and after, the selection pressures were strongly towards higher S and there were few low S barbarians to counter the effects.

The emergence of a higher S civilization

By the time civilization re-emerged there are indications of an ongoing rise in S, continuing for many centuries. In terms of creativity, India in the fifth and sixth centuries AD was midway between the earlier Ganges civilization and the more conventional India of later centuries. While less innovative in philosophy and thought than the old Ganges civilization, it was the classical age of Indian architecture, sculpture, literature, the arts and science. This was the time, for example, when Indians developed the concept of zero, a surprisingly revolutionary idea which forms the basis of our decimal system. In a book written in 499 AD, the astronomer Aryabhata calculated the value of pi and the length of the solar year with remarkably accuracy. He worked out that the earth was a sphere and rotated on its axis, and established the cause of eclipses. All this creativity is an indication of at least moderate infant C.

But S continued to rise—and infant C to fall—as the centuries passed. Instead of leading to a scientific revolution, Aryabhata's theories were opposed by later astronomers, who preferred to compromise with the demands of tradition and religion. India was becoming less open to new ideas. By the tenth century, scientific works such as medical texts were largely commentaries on earlier works, with little reference to empirical knowledge. Astronomy came to be regarded almost as a branch of astrology. The literature of this later period has been described as imitative and pedantic, and an indication of a less-creative society.

Only in the south was this trend less pronounced. There were states with relatively stable boundaries from the sixth to the eighth centuries, and between the tenth and twelfth centuries the south was more innovative in terms of philosophy, religion and trade. But the overall trend was the same.

From the early eleventh century, north India was overrun by waves of invaders, especially Afghans and Turks. The Indians, by now distinctly high S, accepted these foreign rulers with little resistance. The history of Bengal in the late fifteenth and early sixteenth centuries gives a striking example of this process. It started with a revolt by the Abyssinian palace guards, allowing their commander to take the throne. The Abyssinians were then replaced by an adventurer of Arab descent, and then in 1538 by an Afghan nobleman fleeing war further west. The general population accepted these changes with an indifference typical of peoples with high S and reduced local loyalties.

In the sixteenth century most of India was conquered by the Mughals, the descendants of Mongols from the steppes of Asia. After their empire collapsed it was acquired, piece by piece and with almost casual ease, by the British. They overcame several serious challenges to their rule, including the Maratha Wars of 1775–1818, the Sikh Wars of 1845–49 and (the only challenge from within the British territories) the Indian Mutiny of 1857–58, all of which were put down by armies composed largely of native troops. This does not mean that Indians in general were happy with foreign rule, merely that their opposition was not fervent enough to make such rule impossible.

A Continuing Rise in C

Again, as in China, C and V continued to rise even as infant C retreated. An increasingly austere form of Hinduism gained ground, rolling back the Buddhist wave of earlier centuries. Brahmanical Hinduism, far more than Buddhism, lowered the status of women. And subordinating women is the single most effective way of maintaining V and thus C, since it transmits anxiety to their infants (V) and controls their sexual behavior (C and V). Islam, the archetypal high-V religion, served the same function in much of northern India.

India is now clearly a high-S society, with religious and cultural traditions relatively resistant to Western influences and not especially wealthy. The treatment of children is typical of peoples with the high-S genetic factor. For instance, in Rajput communities in northern India, although a busy mother may occasionally leave an infant to cry for a while, in most instances a crying baby will receive instant attention. Once slightly older he is carried so constantly that he gets little chance or opportunity to crawl.[290] This is in sharp contrast to the much sterner, harsher treatment meted out to older children. Similar patterns can be seen in southern India, although punishment of older children may be less harsh.[291]

Greece

The final destination on our tourist's epic journey of the fifth century BC, is ancient Greece. This was the first European society to show signs of high infant C, at about the same time as it appeared in China and India. The Greeks of the sixth to the fourth centuries BC showed all the hallmarks. Their city states were fiercely independent and their civilization creatively brilliant in a way which has scarcely been seen before or since.

Evidence of early control in ancient Greece

In studying ancient China and India we have looked for signs of infant C in nation states, reverence for hereditary rulers, and creativity, but we have no direct evidence of how children were raised. For the Greeks, however, we are more fortunate, and it is clear that they controlled children from a very early age. Spartan nannies were notable for their rigorous treatment of infants, as recorded by Plutarch. Plato also believed that children should be taught to obey from well before the age of three:

> Then until the age of three has been reached by boy or girl, scrupulous and unperfunctory obedience to the instructions just given will be of the first advantage to our infantile charges. At the stage reached by the age of three, and after the ages of four, five, and six, play will be necessary, and we must relax our coddling and inflict punishments—though not such as are degrading.[292]

Aristotle had a similar view, even though he was writing at a time when Greek C was declining and parents less likely to be strict.

> The legislator must mould to his will the frames of newly-born children ... To accustom children to the cold from their earliest years is also an excellent practice ... and children, from their natural warmth, may be easily trained to bear cold. Such care should attend them in the first stage of life.[293]

He believed that children's physical and moral training should commence very early:

> He must also prescribe a physical training for infants and young children. For their moral education the very young should be committed to overseers; these should select the tales which they are told, their associates, the pictures, plays and statues which they see. From five to seven years of age should be the period of preparation for intellectual training.[294]

Like the peoples of China and India at this time, the Greeks had never gone through a major civilization collapse which would raise S by favoring people with lower infant C. They fought off the Persians and thus did not experience a mass migration of high-S peoples into their lands at that time.

Greek creativity and city states—signs of high infant C

The Greeks, in their brief cultural flowering, were to have a huge impact on future ages. They achieved remarkable breakthroughs in philosophy, science, art, architecture and literature. The histories of Herodotus and Thucydides are astonishingly modern in tone, far closer to the modern West than (say) the works of medieval scholastics. The Greeks also created sophisticated political systems, including the first democracies. All this is familiar because their culture influenced Western development, and they had, like us, high infant C (albeit not to the level of nineteenth-century Europeans). Such a character also made possible their remarkable commercial success, with trade routes and colonies all over the Mediterranean.

In terms of creativity Greeks had another advantage: a low level of child V. We have seen that people with high infant C are intensely loyal to leaders similar to themselves in culture and language. Adding low child V, which means the Greeks had little reverence for authority as such, the result was city states and (in many cities) democracy.

The only exception to this was Sparta, which was notoriously different from other Greek states. The Spartans evidently had higher child V due to their exceptionally severe punishment of older children. This made them great soldiers but lacking in creativity. They were notoriously conservative, superstitious and brutal to their subject peoples.

Spartans aside, the city states of Ancient Greece are the model of a dynamic, creative society. The combination of infant C with low child V explains why city state civilizations are creative by their very nature, a phenomenon also seen in northern Italy during the Renaissance and Holland in the sixteenth and early seventeenth centuries. It also explains why the creativity of ancient Greece was at a maximum shortly after the presumed peak of C in the fifth century BC, rather than centuries later as in Rome. But this very openness allowed V and C to drop very quickly, making the Macedonian conquest possible scarcely more than a century later. The Roman character, aided by the rigidity of higher child V, was to prove more enduring.

A moderate rise in S

Collapsing civilizations tend to develop rising S, but for Greece this was modified by relatively large incursions of lower S barbarians when the protection of Rome weakened, notably Slavs from the sixth century AD.[295]

Greece in the twenty-first century is therefore a moderate-S society, with levels of infant C higher than in India or China but below those of northern Europe. Though under Ottoman rule for several centuries, the Greeks rebelled frequently before gaining their independence with Western help in the early nineteenth century. This kind of nationalism is characteristic of high infant C. On the other hand, Greece did not become an industrial powerhouse like Britain or Germany, and the level of corruption and dysfunction in modern Greece suggests infant C well below the levels of northern Europe.

The success of high-S societies

People in the West tend to judge other cultures in terms of two main characteristics: wealth and human rights, the latter often seen as virtually synonymous with democracy. Societies that are poor or autocratic or both tend to be seen as lagging behind the West, a view reflected in terms such as "backward," "third world" and now "developing." Only a few high-S societies in the modern world have achieved Western levels in economic and constitutional terms, these being countries such as Singapore and Taiwan with high C and very low V. India is democratic but poor. China is more affluent but a dictatorship.

But biological success cannot be judged in such ways. Societies that grow and spread and outcompete their neighbors must win out over those which fail to replace themselves, or allow themselves to be overrun. From this viewpoint, the success of the high-S cultures of India and China is beyond question. Not only have they absorbed and assimilated all invaders but their peoples have spread out into many other countries, often becoming commercially successful because of their high C work ethic.

High levels of S do not in themselves prevent people from being commercially successful, provided C is strong enough, and any form of C is associated with hard work and self-discipline. In the modern West where C is dropping rapidly, Indian and Chinese immigrants have done very well, helped by the fact that many are tertiary graduates and thus a relatively elite group.[296] China itself has done well economically in recent decades, aided by the flexibility that arises from an initial drop in V and thus child V, as a result of wealth, urbanization and Western influences. Based on the community study quoted in chapter five, the Chinese seem to start training children somewhat earlier than Arabs, which would make their infant C higher than that of the Middle East though lower than in Japan or Europe. This also helps to explain their outstanding economic success in recent times.

Of course, all of this also means that such people are no longer immune to a catastrophic fall in C, signs of which can be seen in the plummeting birth rates in East Asia. High S makes people more conservative and somewhat less likely to become wealthy, but is no defense against industrial wealth and urban living.

In the next chapter we return to the region which launched our examination of S—the Middle East. Its history gives a better understanding of the development of Judaism, Christianity and Islam, and their crucial influence on the West.

Testing

Detailed studies of parents from different ethnic groups would give a more accurate idea of levels of control and punishment at different ages, to test whether the predicted patterns are found. We would expect populations with the least experience of civilization (Japan, Europe) to control children relatively more in infancy, and those with the most experience of civilization, especially the Arab Middle East, to control children relatively more in late childhood. Chinese should be similar to Arabs but more inclined to early control.

CHAPTER FIFTEEN

THE TRIUMPH OF THE FUNDAMENTALISTS

There is no God but Allah.
—The Quran

Imagine a society where epidemic diseases are common and most children die before the age of five, and yet the birth rate is so high that only famine keeps the population in check. Women are illiterate and confined to the home, their sexuality rigidly controlled. They can be beaten or even killed at the whim of their menfolk. Many gain little pleasure from sex as a result of genital mutilation in childhood. Life for everyone is a grinding struggle for survival, with much of the slender agricultural surplus taken by the rapacious tax agents of an alien and hated power. Religion and tradition are all-pervasive.

This is the traditional culture of the Middle East and it is arguably the most advanced culture on Earth. Most advanced not in terms of technology or wealth but in its ability to endure and reproduce. It is the end product of thousands of years of cultural evolution. In the future, it may well be our world as well.

In chapter fourteen we saw that the early civilizations of China, India and Greece showed unmistakable signs of high infant C. They were highly creative, especially in the field of thought, and formed stable nations and city states which lasted many centuries. In all these cases the decline of C and V led to civilization collapse, followed eventually by the rise of a more stable and conservative culture with low infant C. This change had two crucial elements. One was the development of religions and cultures with ever more powerful V-promoters and C-promoters. And the second was what appears to be a genetic change making parents more indulgent of infants and less indulgent of older children. The effect was to increase the more conservative child V, and to minimize the creative but unstable infant C.

This chapter will follow the same process in the Middle East, where the earliest known civilizations arose and where high S enabled the Eastern half of the Roman Empire to survive while the Western Empire did not. It was a process of cultural evolution played out over thousands of years, and characterized by the development of a succession of major religions—first Judaism, then Christianity, and finally Islam. All of these contained a mixture of C-promoters developed in the settled lands, and V-promoters from the harsher outlying regions.

But the important point is not simply the development of a religion such as might be contained in a holy book, but of an entire complex of behaviors and customs which may be termed "fundamentalist." In Eurasia such customs tend to be associated with Islam, but they began to develop very much earlier.

Sumer and Akkad

At some time in the fourth millennium BC the Sumerians migrated into the region that is now Iraq. All indications are that their civilization they developed had high infant C. Three thousand years before Rome, the Sumerians created splendid city states such as Kish, Nippur, Uruk, Lagash and Ur, which thrived and maintained their independence for more than a thousand years. They were great innovators—another aspect of high infant C. Among much else they were responsible for the first system of writing, the first known codes of law and administration, the potter's wheel, the making of bronze, wheeled vehicles, and the abacus. They were the first to create formulae for calculating the area of a triangle and the volume of a cube. Their fascination with the number 60 is still reflected in how we measure time, as well as the 360 degrees in a circle.

In the late twenty-fourth century BC Sumer was conquered by Sargon of Akkad, and then endured barbarian invasions and a prolonged collapse. It was not until 1763 BC that political unity was restored by Hammurabi of Babylon, a Dark Age of almost five hundred years.

In every respect this is comparable to the early histories of India, China and Greece, though preceding them by two thousand years. A civilization of creative nation or city states is united into a powerful empire, which soon collapses into an extended Dark Age. Our explanation, as given in the past two chapters, is that high infant C made this society brilliant but unstable, leading to a collapse in V and C.

Cultural evolution and the barbarians

All these early civilizations had low S, which means high infant C relative to C. This is no different from the modern West, except that these ancient civilizations had a lower *absolute level* of C. If it were possible to put a number on it we might place the peak C level of most hunter-gatherers at 1, Neolithic farmers at 3, Sumerians at 5, the Greeks and Romans at 7, and nineteenth-century Europeans at 9. Later civilizations had a higher level of C than the Sumerians because they had the advantage of thousands of years of cultural evolution, which involved the development of religions and other traditions with more powerful C-promoters and V-promoters. The key driving force was competition among families and societies, in which people with higher C and V tended to overcome or outbreed those with lower levels.

C-promoters are necessary for the rise of civilization, and civilizations tend to produce more powerful C-promoters with time. Settled peoples have a great deal to gain from higher C: not just larger political units but also more hard-working and productive farmers and merchants. The immense resources the Sumerians poured into their temples and religious systems can be understood as an investment in making their societies more competitive (not that they consciously saw it this way).

V-promoters are different. They tend to be developed most fully by barbarians in the harsher and more unstable environments away from the fertile river valleys. The process of cultural evolution, especially as observed in the Middle East, is one in which C-promoters spread from civilization to the barbarians, while V-promoters are most fully developed by barbarians and brought back to civilization.

This is illustrated by tracing the source of the first barbarian invaders of Sumer. In later ages barbarians emerged from the plateaus and mountains of central Asia (modern Iran and Afghanistan, and the central Asian steppes), and finally from the Arabian desert. But the Akkadians did not. They came from central Mesopotamia, a relatively flat and fertile land. At first sight this is odd. If V is produced in harsh environments with periodic famines, how could the conquerors come from here rather than more inhospitable regions? It was undoubtedly a poorer and more backward area than Sumeria, and may well have experienced famine, but conditions cannot have been as tough as in the mountains and deserts.

We can start by recognizing that a harsh environment alone is not enough to create the highest level of V. It will certainly be increased by occasional famine, especially when interspersed with times of more plentiful food or a sudden increase in prosperity. But maximum V can only be generated when famine is reinforced by cultural V-promoters such as patriarchy and limits on sexual activity.

Also, effective barbarians need organization to attack and overrun wealthy civilizations. Even the most degenerate state can raise thousands of soldiers, many of them experienced professionals. Defeating them requires the ability to form large coalitions at considerable distance from their home bases, among people who might be more accustomed to fighting each other. It also requires warlike ferocity and a willingness to suffer high casualties in a pitched battle, something rarely found among hunter-gatherers and small-scale agriculturalists. In other words, successful barbarians require not only famine but cultural technologies to increase both V *and* C.

The best place to find such technologies, of course, is from the civilized peoples themselves. This is why the first barbarians emerged from areas close to existing civilization rather than from harsher, more remote regions. The Akkadians of ancient Mesopotamia had been close neighbors to the Sumerians for well over a thousand years. Even the barbaric Gutians who took down the Akkadian Empire were from the headwaters of the Diyala River, only about 300 kilometers from the Sumerian border.

We have seen the same pattern in ancient China, which was first unified by the semi-barbarian but agricultural state of Qin in the Wei river valley. Nomad raiders from the north were a problem, encouraging the construction of the defensive works which eventually developed into the Great Wall. But it was to be five hundred years before true nomadic horsemen from the steppes overran China, and 1,400 years before the nomads reached their peak of power under Genghis Khan around 1200 AD.

These cultural technologies arose early in the history of civilization. The laws of Hammurabi, which drew heavily on earlier codes, included provisions to control sexual activity with heavy religious sanctions. For example, a woman accused of adultery had to swear an oath before a god or be drowned in the Euphrates, and the chastity of unmarried girls was strictly protected. The subordination of women was also firmly established by this time. Women were considered the property of their husbands and could even be seized in settlement of a debt.[297]

Without conquest the spread of such cultural technology must be slow, but it happens gradually through trade and migration. The Akkadians, Gutians and other barbarians adopted some aspects of Sumerian religious systems, and as a result became fiercer in war and better organized as their levels of C and V rose. These peoples' V was also raised by the increasingly frequent famines exacerbated by incessant feuds and the absence of central authority, so they were more open to cultural V-promoters which could take their V still higher.

When V and then C began to drop in the neighboring civilizations, as they always must, these fiercer and better-organized barbarians were ready to move in and take over. It is natural for humans, and indeed any primate, to copy the behavior of higher status individuals, and so the conquered peoples began to imitate the high-V invaders and raise their own levels of V. And because V combined with high population density increases stress there will also be a long-term rise in C, since cortisol reduces testosterone.

In other words, what occurs is a two-way process whereby civilized peoples and barbarians reinforce each other's C and V. Civilizations are better at developing C because this is vital for farming. Barbarians are more likely to develop V because of their experience of famine and because it is useful for war. And key behaviors such as sexual restrictions and domination of women support both V and C, so cultural norms that support one tend to support the other.

As already indicated, this is a slow process. People do not easily accept values requiring big changes in behavior. A man who is used to regular sex might accept abstinence for a week, if a priest told him it would cause the gods to bless his crops. He would be less likely to abstain for a month.

A religion promoting high C cannot be fully accepted by people with much lower C, so behavior in the beginning will only be slightly affected. But as levels of C and V rise, so people are willing to accept more stringent codes which further affect behavior in a kind of ratcheting-up effect. The same problem was encountered by early Christian missionaries in northern Europe, for whom nominal conversion of the people was a necessary first step. After that began a centuries-long struggle to reduce or eradicate traditional lower-C behavior, such as the taking of multiple wives.

Israel and the development of Judaism

A vivid illustration of how a cultural technology can move from settled to nomadic peoples and then back to civilization can be found in the Hebrew bible. During the second millennium BC the Jewish people developed a highly advanced cultural technology, including the sexual restriction of both men and women, as well as the belief that there was only one God— an innovative theological concept at the time. The stories and historical records that form the book of Genesis well illustrate the early stages of this process.

The story starts with the figure of Abraham, who appears to be from an Amorite herding society in northern Mesopotamia. The customs described in Genesis, such as the adoption of an heir who had to give way to a "legitimate" son, are familiar from contemporary tablets discovered in this area.[298] And of course, Abraham's later career was that of a successful pastoral leader, not a lifestyle to be taken up easily by an soft city dweller.

At the same time, his association with the Sumerian city of Ur is highly plausible. Though his ancestors had typically Amorite names, the names of Abraham's generation tend to be Sumerian. Sarai (his wife and sister) is the Akkadian name for the wife of Sin, the Sumerian moon god, and Terah's daughter Milcah was almost certainly named after Sin's daughter Malkatu.[299] Abraham thus represented an Amorite herder who lived for a time in Sumer and was influenced by their culture, itself based heavily on their advanced religious traditions.

The same process is seen even more powerfully in the Exodus story. The Israelites clearly kept much of their pastoral tradition and culture, as indicated by a very high birth rate, but were also influenced by the sophisticated and advanced Egyptian culture. It is significant that Moses, the great lawgiver, was brought up as an Egyptian prince and would thus have been both literate and familiar with Egyptian laws. Laws are a key aspect of a high-C culture, and religious laws are powerful and effective C-promoters.

Thus it was that the Israelites who moved into Canaan had a religious and cultural system combining effective C-promoters and V-promoters. With relatively high C, and V levels elevated by their harsh desert experience, they were both aggressive and cohesive enough to displace native peoples and eventually take over the land. By the time of the Babylonian exile in the sixth century BC they had developed still more powerful C- and V-promoting traditions, strong enough to survive transplantation of the elite to another country and eventually to endure two thousand years of exile from their homeland.

The Ten Commandments provide a sense of why this religion was so effective. Three of them (keeping the Sabbath, respect for parents and forbidding adultery) directly support V and C. Another three (only one God, no idols, no blasphemy) build reverence for a single deity and no other, an increasingly important principle as the demands of the Jewish faith grew in rigor. Polytheists can escape from overly demanding deities while maintaining their basic worldview, while monotheists, putting their whole faith in a single God, cannot. The other commandments prohibit violence, theft, dishonesty and greed, important for maintaining social cohesion and also C-promoters (in the sense that any control on behavior elevates C).

Combined with a host of other laws given divine sanction, and with inspiring and colourful narratives of great men and their deeds, Judaism formed a powerful and effective tradition that was the end-product of thousands of years of cultural evolution. Two thousand years ago these advanced Jewish traditions began spreading to other peoples through Christianity and then Islam, both of which were powerfully influenced by Judaism. And through Christianity these traditions have largely built the modern world, by creating the temperament that has made advanced industrial civilization possible.

The strengthening of C-promoters and V-promoters

Even before this, cultural technologies developed first in Egypt and Iraq spread gradually to the surrounding peoples. As they did so, and as C-promoters gradually became more powerful and effective, the size of empires grew (see Fig. 15.1 below).

This was helped by the fact that during the long Dark Age that followed the collapse of Sumerian civilization, the peoples of Mesopotamia had clearly evolved from low to high S. From the time of Hammurabi onward there were no more nation or city states in the Middle East but increasingly large and cosmopolitan empires. Nor did any civilization show the innovative brilliance of the Sumerians, despite much greater wealth and larger populations.

Fig. 15.1. Expansion of civilization in the Middle East.[300] Civilization expanded steadily from its Sumerian origins to successively harsher terrain as C- and V-promoting technologies were taken up by barbarians.

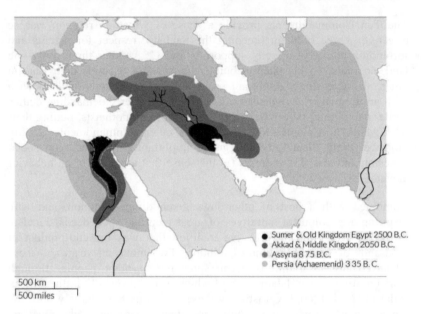

Sumer & Old Kingdom Egypt 2500 B.C.
Akkad & Middle Kingdon 2050 B.C.
Assyria 8 75 B.C.
Persia (Achaemenid) 3 35 B. C.

500 km
500 miles

The reason, as in India and China, is that the invading tribes were pastoralists and much less numerous than the dense populations of the river valleys. This was even more the case in Egypt, which was largely surrounded by desert with very few inhabitants. Nor did Mesopotamia suffer depopulation as in the Roman Empire, where the wealth of conquest and vast slave estates allowed impoverished farmers to drift to the cities.

Thus the nomadic invaders had a limited impact on the gene pool, but a *major* impact on culture by strengthening V-forming traditions. Especially influential were the Aramaeans, a group of nomads who invaded Mesopotamia in the twelfth and eleventh centuries BC. Despite the fact that they never formed a major empire and were soon crushed by Assyria, their language became the *lingua franca* of the Middle East. Their very lack of unity, combined with military ferocity and what seems to have been a fast-growing population, suggests an unusually high level of V, even for a nomadic people.[301] To us it seems strange that people should adopt the language of illiterate pastoralists but reject that of the culturally more advanced Greeks when they took over in the late fourth century BC.

It becomes less so when we understand that the high-V culture of the Aramaeans was more beneficial to survival and success than the declining-V culture of the Greeks.

And V-promoters continued to strengthen, as indicated by the growing subjection of women. Repression of women limits their sexual activity, and also makes them chronically anxious, both of which support V. This is first evident in Babylonian times around 2000 BC, when women were expected to cover their bodies and faces and be chaperoned in public. The Assyrians took this further by insisting they stay home most of the time, concealed behind curtains, a custom also adopted by the Persians.

The subjugation of women is not only inhumane but expensive. It takes a lot of effort to seclude women and reduces their contribution to agriculture, industry and the economy in general. But the military success of the Assyrians and Persians was ample reward for the cost involved in segregating their women. In the long term, biological and cultural success is based not on wealth or even happiness but on the number of surviving children and the status they hold. Military prowess, in both offense and defense, is an effective way to achieve such success.

Islam

All these trends culminated in the culture associated with Islam, which arose in the harshest and most inhospitable environment of all—the deserts of Arabia. Here again, as with the Israelites in earlier times, can be seen the influence of cultural technologies from more civilized lands. There were sizeable Jewish and Christian communities in Arabia at the time. According to Islamic tradition Muhammad accepted the Jewish scriptures, recognized Jesus as a prophet, and established Ramadan as a pillar of the new Muslim faith. Non-Muslims would suggest he was influenced by the Lenten fast of Christianity, but gave it a far more rigorous form by forbidding food and drink during daylight hours. Unlike shorter-term and more frequent fasting which would tend to support C, Ramadan has the physiological impact of an occasional famine and is thus a uniquely powerful V-promoter.

Psychologically, the Islamic stress on absolute submission to Allah suits a temperament which has high child V and a great deal of stress as a result of severe punishments in childhood. It is thus receptive to powerful authority. The word "Islam" means "submission", though of course in the sense of submission to God. Taking on such an attitude probably also increases V, given the effect feedback cycle, whereby any behavior resulting from C or V tends to increase C or V. Combined with exceptionally high V from a harsh desert environment, this helped make Islam the ultimate high-V religion. V-promoters are key elements of Christianity and Judaism, but to a lesser extent.

There is far more to the Islamic way of life than fasting and segregating women, of course. Praying five times a day acts to increase C, as does avoiding alcohol. The custom of eating with the right hand, leaving the left for ablutions, is an effective hygiene measure in a culture without modern sanitation. And this is only one of many health measures associated with Islam, such as ritual washing. Then there is the Qur'an itself and the sonorous power of the Arabic language, with an attractive system of ethics including a focus on alms-giving and the equality of believers. Putting all this together created a powerful religious technology which made its followers more aggressive, confident, united and with a higher birth rate than any competing civilization.

It is not surprising that the united Arab tribes should sweep over the Middle East in scarcely more time than it had taken Alexander a thousand years earlier. The tribes of desert and steppe make fierce warriors once their harsh living conditions combine with the cultural technology for maintaining higher V. The combination of V and stress also makes them better able to accept authority and thus unite in larger groups, but against this must be set the incessant feuding and disunity usually associated with high V. However, once united they are almost unstoppable. What caused the Arabs to unite in the seventh century is almost certainly a lemming cycle G period, which is associated both with unification and military expansion after the peak. The period before the G period is also relatively open to new ideas, which would have made the Arabs more open to the radical new principles contained in Islam.[302]

The most striking consequence of the early Islamic conquests is what came after, with the conquered peoples largely adopting the religion, dress and even the language of the Arabs. Unlike in later times, this was not a matter of deliberate policy. The Caliph 'Umar (634–44 AD) is alleged to have accepted the surrender of the Syrian Christians on the condition that (among other things) they must *not* adopt Arab dress or speech. The Arabs were concerned to maintain their distinct identity as conquerors, not to mould their subject peoples to match them.

Another reason for discouraging conversion to Islam was economic. A tax was levied on non-believers, and the Arab rulers were reluctant to let it go. So for some eighty years they continued to levy it on converts. Only very gradually did the conversion of non-believers become accepted, and it took still longer for it to become an aim of Muslim policy.

But the reason the subject peoples chose to adopt Arabic culture is obvious if we follow the lessons of history. High V is advantageous to settled peoples. It makes them confident, aggressive and with a high birth rate. Those who adopted high-V customs, in imitation of their new overlords, would outbreed and outcompete those who did not. Muslims also adopted existing cultural technologies such as the custom of secluding women. This was not a Bedouin practice but was common in the cities of the Middle East.

In pre-Islamic times the seclusion of women seems to have been confined to certain classes. The effect of Islam has been to spread the custom throughout society. This is part of the process by which Islamic culture has become more rigid and severe, and thus more effective in the one way that matters for long-term success—having more surviving children. This is again a form of ratcheting effect, by which high V customs increase V, which makes even more extreme customs possible.

Fundamentalist versus liberal Islam

Islam in the modern world takes many different forms. The dominant tradition in Indonesia, for example, is relatively mild. One of the most severe forms is Wahhabism, which advocates extreme separation of the sexes. It is no coincidence that this very high-V version of Islam became dominant in Saudi Arabia, the area of the Arab world which until recently has had the harshest climate and living conditions.

A pre-Islamic custom taken up by Muslims in many areas was female circumcision, a method of reducing women's sexuality. The effect of this would be to increase C and V.

It has often been said that such practices do not belong to the "true" religion of Islam. For example, Muhammad seems to have had a relatively positive relationship with women, listening to his wives and helping them with household chores.[303] But what is important when it comes to cultural success or failure is not the idealized religion of a sacred text but how people behave. This is the product of how that text relates to and combines with folk customs and pre-Muslim traditions—what we have referred to as "fundamentalism." Put together, these have a massive effect on temperament and behavior.

People reared in this way think, feel and react quite differently from those brought up in the West, or in the more liberal culture of urban Muslims. Failure to recognize this fact has had disastrous consequences. The costly, chaotic occupations of Iraq and Afghanistan by the United States and its allies arose from a belief that human beings are fundamentally the same and that removal of tyrannical regimes, plus some education, will turn these countries into free, prosperous democracies. But five thousand years of cultural evolution have ensured that people are *not* the same, and political regimes tend to reflect these differences.

Parliamentary democracy in the developed nations of the West is intimately tied to the relatively high infant C and very low child V and stress found in those countries, resulting in national loyalties and impersonal political systems. Arabs, Persians, Afghans and the other Muslim ethnic groups of the Middle East and Central Asia are epigenetically primed by centuries of cultural evolution for personal loyalties rather than impersonal institutions. High child V and stress makes them amenable to harsh authority and inclines them to rebel against rulers they perceive as weak. It also makes them rigidly conservative, so that education of women, seen by Westerners as an unalloyed good, is seen by many fundamentalist Muslims (quite correctly) as a threat to their culture and way of life.

Muslims have not been immune to the degenerative effects on C and V of wealth and Western values. The birth rate of many Muslim countries has dropped dramatically in recent decades, as has that of Muslim immigrants into Western countries. But they have proved more resistant to decline than any other culture, and in rural areas and countries such as Yemen and Afghanistan, the traditional ways remain very strong.

People in the West see the traditional culture of the Muslim Middle East as primitive and "backward," and there are constant calls for modernization. In fact, as had been seen, Islamic culture is anything but backward. Civilization first arose in Egypt, Mesopotamia and the Indus Valley in what is now Pakistan. It is no coincidence that these lands, with the longest experience of civilization, are now strongly and fervently Muslim. Long experience of civilization has bred a high-S genotype and culture which perfectly adapt people to survive and expand their numbers in dense agricultural and urban populations.

Such countries tend to be poor (if we leave out the anomalous effects of oil wealth), since their peoples lack the temperament for industrialization. But wealth at that level is of no benefit in the long-term struggle for survival and success. To paraphrase Christian scripture, what does it benefit a civilization if it gains wealth but loses its strength and vigor? The advantages of Islam can be clearly seen in countries with mixed populations. Lebanon once had a Christian majority but is now 54% Muslim.[304]   In Communist Yugoslavia the provinces with Muslim populations grew much faster and received tax revenue from the wealthier Christian states. The population of Kosovo, the spiritual homeland of Christian Serbia, grew from 733,000 in 1948 to over two million in 1994, with the Muslim component surging from 68% to 90%, and lately going even higher.[305]

Meanwhile, Muslims are migrating into Europe where C is in decline, the birth rate is far below replacement level, and people no longer have much faith in their own culture. Over the next few decades, as the next chapter will indicate, the native peoples of the West will become feebler and fewer. This means that on current trends Europe will become an Islamic continent in a century or so. The 1,400-year struggle between Islam and the West is coming to end.

Testing

Traditional Muslims, and especially Fundamentalists and Wahhabis, should have higher levels of V, child V and stress, as indicated by epigenetic and hormonal tests, than people from other groups.

CHAPTER SIXTEEN

THE DECLINE OF THE WEST

Men are not punished for their sins, but by them.
—Elbert Hubbard

We live in a golden age.

There have been no major wars in Europe or North America for nearly three quarters of a century, and only minor regional conflicts elsewhere. The continuing magic of the machine age has brought prosperity to billions. In most countries epidemic disease has been curbed and new-born infants can expect a long and healthy life. Democratic forms of government continue to spread, albeit unevenly, around the globe. The population bomb has been all but defused. Science and the arts are flourishing, including brilliant new art forms such as movies and computer games that previous societies could not have imagined.

Conventional wisdom is that the West has reached a new equilibrium, that birth rates plus immigration will maintain populations, that Western nations will remain peaceful but capable of defending themselves, and that they will remain open and tolerant liberal democracies. An example of this thinking can be found in Francis Fukuyama's book *The End of History*:

> What we may be witnessing is not just the end of the Cold War, or the passing of a particular period of postwar history, but the end of history as such; that is, the end point of mankind's ideological evolution and the universalization of Western liberal democracy as the final form of human government.[306]

Such concerns as there are mainly relate to climate change—the idea that increasing carbon dioxide in the air will gradually warm the Earth and cause the oceans to rise.

In previous chapters we have seen history in a very different light—not as a steady progress to its current peak but as a pattern in which creative and wealthy civilizations collapse and are replaced by more conservative and stable ones. This tells us that there is a far greater disaster facing the West than the worst possible scenarios for climate change. Western civilization is in a decline which has been in motion since the late nineteenth century, and has accelerated greatly since the 1960s. This is the same decline as occurred in countless other civilizations, from Sumer to ancient China and from India to Rome, the end point of which is complete social and economic collapse. But for the first time, if biohistory is correct, we may now understand exactly *what* is happening, and *why*.

Chapters six and seven examined the reasons for the rise of the West, and in particular the workings of the civilization cycle. It is now time to follow this cycle into the twentieth and twenty-first centuries.

The decline of V, child V and stress

The rise of the West followed the model of the civilization cycle in which high levels of V and stress helped drive C towards an unprecedented peak in the mid-nineteenth century. By this time V and stress had already been in decline for several centuries. As noted in chapter seven, people in the nineteenth century were far more humane in their outlook and behavior than their sixteenth-century ancestors had been. Prisoners were no longer tortured, and death sentences were increasingly reserved for the most serious crimes. Women were acquiring greater rights and children were less brutally punished. Governments became less arbitrary and (in many countries) more democratic.

As V and stress continued to fall during the twentieth century, all these trends gathered pace. Sweden executed its last prisoner in 1910, Switzerland in 1940, Germany in 1949, Britain in 1964, and France in 1977. The United States almost halted executions in the late 1960s, though with a small resurgence in recent decades, and today it is the only Western country to impose it.[307]

Physical punishment of children has shown similar trends. It was officially banned in schools in the Netherlands in 1920, Italy in 1928, Norway in 1936, Austria in 1974 and in Germany through the 1970s and 1980s.[308] In Britain, caning was officially banned in state schools and some private schools in 1987, followed by a complete ban in the remaining private schools in 2003.[309] In Japan it was officially banned in 1947.[310] The United States has banned it in some states but not others.

The status of women has risen dramatically since the nineteenth century as the result of a continuing decline in V. In 1893 New Zealand gave women the vote, followed by Australia in 1902, then Finland in 1906 and Norway soon after.[311] Britain gave equal voting rights to both sexes in 1928. By the 1920s women had received the vote in Denmark, the United States, Austria, Germany, Canada and the Netherlands, although France did not follow suit until 1944 and Switzerland not until 1971.[312] And as the twentieth century passed, women gained ever greater social and economic freedoms, emerging from the home and establishing themselves in professional, industrial and military spheres.

The rise of democracy also indicates falling stress and a decline in child V. The totalitarian regimes of Germany and Italy gave way to stable democracies after 1945. The last holdout in Western Europe was Spain, which made the same transition after the death of Franco in 1975.

All these changes indicate that, as predicted by the civilization cycle, V continued to fall in the twentieth century. It should be noted that American V remains significantly higher than that of Europe. As well as the maintenance of more severe punishments, this is reflected in a continuing willingness to fight wars in Vietnam, Iraq and Afghanistan. The same can be said of Russia which permits the physical punishment of children and even soldiers, is readier to fight wars, and has a more authoritarian government - a sign of higher child V.

The decline of V was especially rapid in the nineteenth and early twentieth centuries, with the most significant visible effects taking place by the 1960s and 1970s. This too is predicted by the theory of the civilization cycle, in which an initial fall in V is followed by a much larger decline in C (see Fig. 16.1 below).

But while these changes are benign, others might cause some concern.

Population growth is affected by both V and C. High birth rates are found among peoples who have high V, such as the tribes of deserts and mountains which raided and conquered settled lands throughout history. Population growth is also associated with high levels of C in the civilization cycle. When wealthy urban populations lose V and C, as they did in the late Roman Republic, the birth rate falls. In the end, such populations typically fail to replace themselves.

Fig. 16.1. Civilization cycle in twentieth-century Europe and America. Prior to the 1960s there was a rapid decline in V and stress but only a minor decline in C. After this, C began to decline much faster.

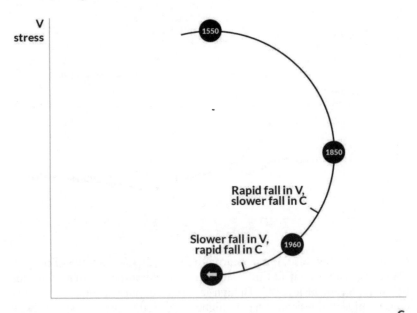

Population decline

After very high growth rates in the nineteenth century, Western countries experienced a sharp decline in fertility in the early decades of the twentieth. Growth picked up again somewhat during the 1950s "baby boom" due to the short-term rise in C caused by the Great Depression (see chapter ten). This resurgence, plus continued growth in the Third World, caused considerable alarm. But concerns about overpopulation have proved unwarranted, at least in the West. On latest figures not one Western country apart from Israel is producing the minimum 2.1 children per woman needed to maintain itself (2.1 rather than 2 to allow for early deaths). In Europe as a whole birth rates have gone from well above to well below replacement in less than thirty years (see Fig. 16.2 below).

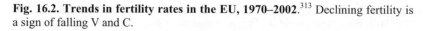

Fig. 16.2. Trends in fertility rates in the EU, 1970–2002.[313] Declining fertility is a sign of falling V and C.

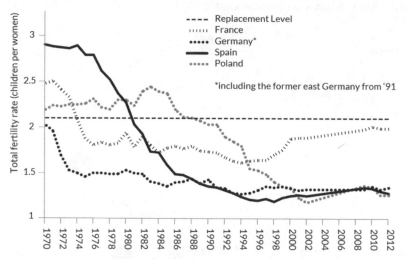

On this indication also, the US currently has higher V than most European countries, with a total fertility rate of 2.01 children per woman, only just below replacement level.[314] Though it might be more accurate to say that conservative Americans have higher V; red states have growing populations, with Utah showing a fertility rate of 2.45 (presumably the Mormon influence) and Alaska 2.28, both safely above replacement level. Blue states are more European. Rhode Island has a fertility rate of 1.5, well below replacement level, and Washington DC is a demographic sinkhole at 1.11.

East Asia is following the same path. Without immigration, the Japanese population is expected to decline by almost two thirds over the next century (see Fig. 16.3 below).

On current trends, European populations can only be maintained by massive immigration from poorer countries, especially the Muslim Middle East and North Africa. And this is based on the optimistic assumption that birth rates will maintain themselves over the next century. Biohistory predicts they will not.

Fig. 16.3. Japanese population: historic and projected.[315] As V continues to fall, the Japanese population will cease to grow and start declining. Only large-scale immigration from poorer countries will present this happening throughout the West.

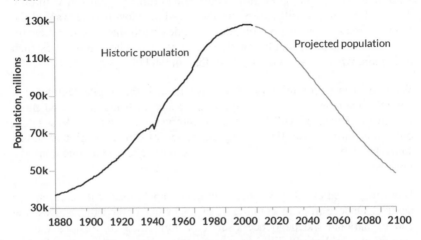

Efforts to halt this decline with financial incentives have proved futile, since the causes are psychological and physiological rather than economic. Of all European leaders only Vladimir Putin has a policy even pointing in the right direction, which is to support the traditional values of the Orthodox Church. Russian fertility is below that of the U.S. but has risen steadily from a low point in the 1990s.[316] However, biohistory indicates that even this will be futile in the long-term, and the policy of most Western governments is in the opposite direction.

Pacifism

After a surge of aggression in the early twentieth century, for reasons discussed in chapter nine, falling V in the West has caused a dramatic decline in the taste for military action. In the United States, a watershed moment was the loss of Vietnam. America's casualties were less than a tenth of what they had been in the Second World War, yet public opinion turned overwhelmingly against the war and forced a staged withdrawal. Pacifism was strongest among young radicals but had spread widely by the late 1960s. War is costly and needs vigorous support to be sustained. The public as a whole was no longer willing to provide such support.

Opposition to war had become even more marked by the turn of the century. In 2003 the invasion of Iraq found majority (but far from unanimous) support in the US, and overwhelming opposition in Europe. Four years later, opinion even in America had turned against it, with only 40% of US citizens still approving the decision. Most people wanted the troops home as soon as possible, despite a death toll that was significantly less than in Vietnam. Public opinion was even more hostile in allies such as Britain, where V was ebbing faster than in the US.

Western distaste for military adventures is such that rogue regimes can commit genocide virtually at will, as happened in Kampuchea and more recently in Rwanda, with little fear of intervention. And the West could only look on in frustration as rag-tag bands of Somalian pirates seized ships from one of the world's busiest waterways, demanding massive ransoms.

Biohistory predicts that Western military capabilities will continue to decline. Once European countries can no longer defend themselves, the end of national independence cannot be long delayed. Just as Greece submitted meekly to Phillip of Macedon in the late fourth century BC, so must the nations of the West submit to any serious aggressor by the end of this century.

Loss of cultural confidence and morale

Another effect of falling V is a decline in morale and cultural confidence. 150 years earlier, Europeans had little doubt that their progressive, industrial and Christian culture was beyond compare. Their colonial empires ruled most of the world, while science and technology progressed at an unprecedented rate. Charles Kingsley mirrored the immense confidence of the time by writing of: "the glorious work which God seems to have laid on the English race, to replenish the earth and subdue it."[317]

Contrast that bold, vigorous confidence with the self-doubt which has grown since the 1960s. To modern eyes, Kingsley's words seem hubristic and arrogant. There is widespread pessimism about the environment, politics and even the value of science.[318] Perhaps the strongest expression of this change is the theory and practice of multiculturalism. As late as the 1970s, most American colleges required history majors to complete a course in Western Civilization. By 2011, only 2% of General Education Programs retained this as a required course.[319] Many people in the West no longer believe their culture is better than any other, in marked contrast to the growing confidence of the Muslim world.

Evidence and effects of declining C

While declining V has disturbing implications, a far more serious issue is declining C. We have traced the rise of C in England and Japan to a peak in the nineteenth and early twentieth centuries, and seen a similar pattern in the Roman Republic. For Rome the fall in C, driven by prosperity and declining V, was disastrous. It ended the Republic, undermined the economy and caused the Empire to collapse. In Rome we saw the civilization cycle played out in full. The West is at a much earlier stage, but biohistory predicts a similar outcome.

Changes to Marriage and the Family

C is a physiological system that allows animals to respond effectively to chronically limited food. When food is plentiful C declines, with both psychological and physiological effects. One of these is that animals start breeding earlier, which tends to increase the birth rate.

In humans, as we have seen, the ability to control reproduction means that falling C *reduces* births, but the underlying biology is the same and it is expressed in the falling age of puberty. This is a feature of all Western countries in the past century and a half (see Fig. 16.4 below).

Popular opinion is that this is the result of better nutrition and thus a 'good thing', but this does not explain why puberty became so *delayed* in the centuries leading up to the Industrial Revolution.

Another characteristic of lower C, in animals as well as people, is that sexual activity starts earlier and is more frequent. Prior to the nineteenth century rising C was reflected in a later age of marriage. Since then, falling C has caused people to advance not so much marriage but the onset of sexual activity.

Kinsey found that women coming of age in the 1920s were already somewhat more sexually active than their mothers, but this trend accelerated from the 1960s. In Britain today, 27% of young women are sexually active before the age of consent, compared with just 4% for those born in the 1950s.[320] In the US, the percentage of females having sex by the age of 17 almost doubled from 20% to a little under 40% between

1972 and 1987. The average age of first sex in the US was 16.9 in 2005.[321] This is further reflected in the surge of extramarital births (see Fig. 16.5 below).

Fig. 16.4. Declining age of menarche in Europe and the US.[322] A declining age of puberty (measured here by age of menarche) reflects falling C as much as increased prosperity.

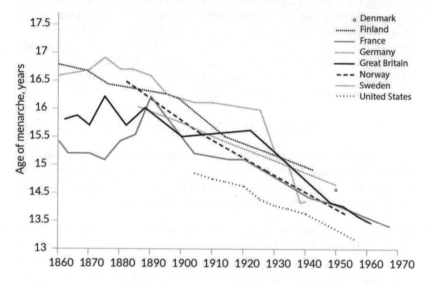

As to why the 1960s were a turning point, this can be explained by the pattern of the civilization cycle. The early stages of declining C show a gradual fall in C with a more rapid fall in V. At a certain point, in and around the 1960s, the fall of V slows and that of C accelerates (see Fig. 16.1)

This represents a radical breakdown of the traditional sexual morality which supported C and raised Western civilization to its greatest peak.

Fig. 16.5. Rates of extramarital birth, England and Wales, 1900–2002: percentage of live births outside marriage.[323]

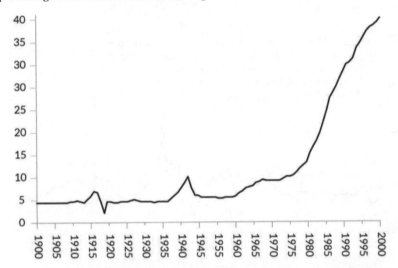

The breakdown of sexual morality is also reflected in surging levels of divorce. Once again, the biggest change has been seen since the 1960s (see Fig. 16.6 below). But divorce statistics actually understate the change since they take no account of irregular unions which are often short-term. A more meaningful measure is the increase in one-parent families. The 1960 US Census showed 9% of children dependent on a single parent. By 2000, 28% of children were dependent on a single parent. This represents a dramatic breakdown in the nuclear family which is the essence of high C.

Fig. 16.6. Marriage and divorce in the United States; 1860–2000.[324] Increasing divorce is a sign of the breakdown in lifelong monogamous relationships, a clear indication of falling C.

In a strange irony, the only European country making any attempt to resist this trend is formerly Communist Russia.[325]

Changes in childrearing

Another and related trend is that children are much less controlled than in previous generations. This is a different factor from punishment, which has been in decline since the sixteenth century. From all accounts, as indicated in chapter six, control of children reached an all-time high in the nineteenth century.

Changing behavior can as usual be reconstructed from parental advice books, since people tend to buy the books that reflect their attitudes. These began advocating leniency in childrearing from the 1920s. From this time there was a gradual rejection of the mechanistic and behaviorist idea of childrearing, and an increasing emphasis on the importance of meeting a child's emotional needs. For example, in 1936 Charles Aldrich's *Babies Are Human Beings* advocated greater leniency and reduced control.[326] By mid-century, Dr Spock took a far more radical position. He argues that the child's emotional needs came before those of parents, imploring mothers to enjoy their children and reject traditional attitudes. In recent decades this "revolution against patriarchy" has almost banished discipline and control from most American families.[327]

Parental control is crucial because it is the bridge between generations, the way in which the C of parents is transmitted to the children. It also slows the process of declining C over generations. People who experience a fall in C, as a result of prosperity and more liberal attitudes to sex, nevertheless have much higher levels of C (and especially infant C) set in childhood. This not only gives them economic advantages but inclines them to exercise significant control of their children, who thus have higher C in their turn. This means that the process of declining C may take some generations.

Indeed, one reason why the link between sexual liberation and social decline has been hidden is that the effects of teenage sexual behavior (for example) may not fully be felt until those affected have children, who grow up and reach positions of leadership. This could be as much as a half century later. In the West, these are the children born from about the 1960s onward.

Changes in moral values and the rise of passive welfare

In tracing the rise of the West we saw how moral attitudes shifted from a focus on personal charity to such values as integrity, self-discipline, hard work and *principled* philanthropy which tried to reform the character of the poor. These attitudes reached a peak in the nineteenth century and especially the Victorian era, and are a key indication of high C.

150 years later, those standards have been all but abandoned. A recent study found that a group of young adults, regardless of social background, had few absolute standards of any kind. The only "traditional" values retained were that murder, rape and robbery were seen as wrong. Otherwise, morality was seen in terms of what one's peers would think and the chance of getting caught, and standards of right and wrong were a matter for personal choice.[328]

As moral standards have declined, the business of government handouts and unconditional charity has increased, free from the stringent moral requirements of the Victorian era. As C declines, governments spend more and more of the national wealth on subsidizing low-income earners, paying state pensions and generally supporting the needy. This can be plotted by the increase in welfare spending over the past sixty years (see Fig. 16.7 below).

Fig. 16.7. Welfare and pension spending as percentages of GDP in the US.[329]

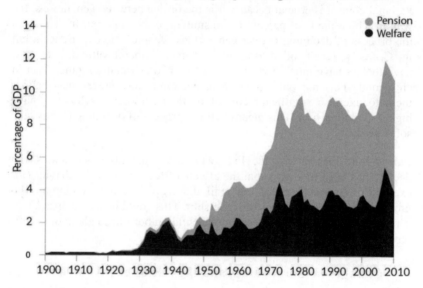

According to the latest census data, 49% of Americans now receive benefits from one or more government programs for which very many do no work.[330] This is a mirror of the late Roman Republic, where the descendants of hard-working farmers enjoyed corn doles and popular entertainment known as "bread and circuses".

Decline of the market; rise of the state

One benefit of high infant C is a strong market economy. In chapter six we saw how non-market institutions of the Middle Ages, such as serfdom and craft guilds, gave way to free markets with minimal government control by the nineteenth century.

An indication of falling C since then is the steady rise of government spending, which is in essence a non-market distribution system. In most European countries this is now well over 50% of GDP.[331] Even in America, behind the trend in so many ways, the government's share has risen from 7% in the early 1900s to more than 42% in a little over a century (see Fig. 16.8).

Fig. 16.8. Total (Federal, State, and Local) US government Spending.[332] Lower C people want governments to redistribute income and are less self-reliant and market oriented. Similar trends can be seen in the late Roman Republic and during the Empire.

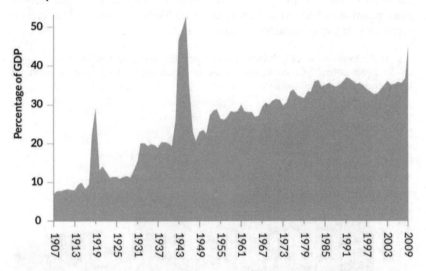

Equally significant is a massive increase in government regulation, with a multitude of laws covering wages, hiring and firing, licensing, planning permissions, rent controls, environmental regulations and more. People with lower infant C are uncomfortable with the market. They believe that government intervention can drive growth, even though practical experience (as, for example, the Communist experiment) suggests otherwise. The same belief in regulation was seen in the later Roman Empire, where Emperors such as Diocletian tried to stem economic decline by tying people to their jobs.

In this as in other areas, Americans are following the trend but somewhat behind other countries. Many Americans are suspicious of high government spending and increased regulation, especially in more conservative states such as Utah and Texas.

Declining work ethic

In the long term national wealth depends on hard work, and the number of hours worked is dropping rapidly with the fall in C (see Fig. 16.9 below). Shorter working hours are often attributed to affluence, which means that people no longer *need* to work so hard. But affluence is not a good explanation for this change. We have seen that higher C people have a more positive attitude to work and are more likely to enjoy it. This can be seen most clearly in attitudes to work.

Fig. 16.9. Average weekly hours worked per US male, all racial groups, aged 25–54 1900–2005.[333] Lower C people are less inclined to work hard, so working hours drop with the fall in C.

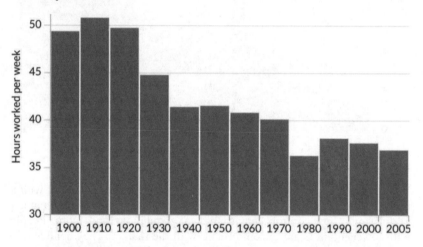

This pattern can be most clearly seen in the young, who have lower C than their elders. A recent report concluded that the poor employment outcomes for younger Britons could be explained by them lacking 'employability' skills such as literacy, numeracy, self-management and a positive approach to work.[334]

Increased debt

Until 1994, 58% of Americans said the most important thing they looked for was a sense that their work was important and gave them a feeling of accomplishment. By 2006, the proportion putting this in first place had dropped to 43%. Until 1994 the reasons *least* likely to be given as first priority were short working hours (4%) and no danger of being fired (6%). But in 2006, 9% gave short working hours as their first priority, and 12% gave job security.[335] Work has become less valued for its own sake.

The economic rise of the West was driven, at least in part, by hard work and a willingness to defer immediate gratification for future benefit. People were inclined toward self-improvement and had a tendency to save and invest rather than spend.

Falling C has the opposite effect. People are now more likely to favor pleasure and comfort over future benefit. At a moderate level this could mean lower levels of savings and investment. At higher levels it increases debt. Thus it is no surprise that debt levels have grown dramatically in recent decades as C has fallen. In the past six decades, debt has grown from 20% to 120% of personal disposable income.

The same attitude impacts on nations, with citizens wanting more government services but not being willing to pay for them. The result is spiraling levels of government debt (see Fig. 16.10 below).

Fig. 16.10. Gross Debt as a percentage of GDP.[336] As C declines and people prefer current consumption to future benefit, they also pressure governments to plunge into debt.

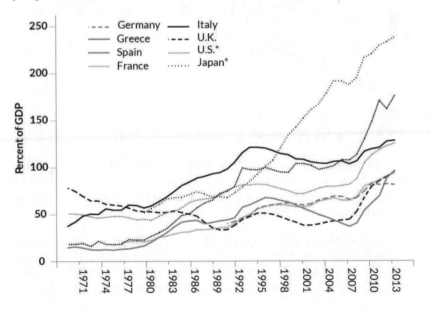

Machine skills, engineering and science

People with moderate C can be competent at jobs requiring personal skills, such as government bureaucracy or retail, especially given low stress. But machine skills and engineering are especially the province of people with high infant C. Continuing the graph of innovations in science and technology from chapter seven, and comparing against population growth, we see that the dramatic rise in the nineteenth century has been followed by an equally dramatic fall (see Fig. 16.11 below).

Fig. 16.11. Key innovations in science and technology plotted against population growth.[337] This is based on 8,583 key scientific and technological innovations, plotted against world population. Science and technology skills were strongest when the West reached its peak of infant C in the nineteenth century, and have since declined.

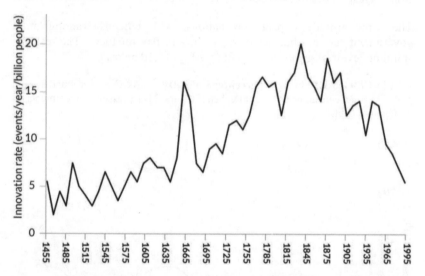

There has been a loss of interest in science, which had brought massive benefits in the nineteenth and early twentieth centuries. The decline of science, in particular physics and chemistry, has been particularly notable in Britain over the past fifty years.

Newspapers have cited a "sustained national decline in demand for physics education" and the Institute of Physics, the professional body for physicists in the United Kingdom, described the trends as "enormously worrying" and with serious implications for national prosperity.[338] There is now also a serious shortage of skilled computer programmers, allowing them to command very high salaries.[339]

American predominance in science and engineering continued after Britain went into decline, but the same trend has now occurred there. The number of engineering doctorates in the US dropped from around 5,700 in 1993 to around 5,330 in 2000.[340] Yet, despite this gradual decline, the US remains the world leader in innovation. One key reason is the American talent for recruiting professionals from higher C cultures such as India and China, although these lack the high infant C of the West at its peak.

Declining nationalism and engagement with the state

People with high infant C are loyal to the nation state and adamantly resist alien rulers. After a last burst of aggressive nationalism in the early twentieth century, Western nations have seen a steep drop in national pride. Confidence in national institutions such as parliaments and political parties has also gone downhill.[341] This is reflected in declining membership of political parties and lower turnout at elections.

Reduced nationalism has helped make possible the rise of the European Union. Not only trade and immigration policy but a host of rules and regulations are now decided by nations acting in concert—a remarkable change from the fierce national rivalries of less than a century ago.

Obesity, infertility, falling testosterone

There are several other changes in recent decades that can be explained by falling C. One is obesity, which has reached epidemic proportions in most Western nations. Obesity has been linked to risk-taking in men and impulsivity in women, single parenthood and poverty, all indications of low C.[342] Overeating is both a cause of lower C and a consequence of it, in the sense of reflecting a lack of self-discipline.

Infertility is also increasing, another indication of falling C. One of the most significant findings from our laboratory research was that sperm count is significantly higher in rats whose mothers experienced calorie restriction in infancy. In humans, sperm count has been declining in Western men over the past two generations. A recent large-scale study found a 32.2% decline in French sperm count between 1989 and 2005, with the proportion of properly formed sperm falling from 60.9% to 52.8%.[343] Falling C is very likely the reason, especially given evidence cited in earlier chapters of lower fertility in other declining C groups, such as the Romans of the second century and decayed Chinese gentry families. Increased infertility will further accelerate the population decline discussed earlier.

Another key finding of our research is that early food restriction can increase testosterone. This explains why testosterone levels of American males have been declining over the past two decades, something to be expected as infant C falls to a lower level.[344]

Disease

Another consequence of falling C is that infectious disease becomes more virulent. Rat experiments show that food restriction (high C) increases resistance to disease, a relationship borne out by the decline in disease in Europe as C rose. The Black Death of the late fourteenth and early fifteenth centuries occurred during a lemming cycle G-150 period (when populations are less resistant to disease), but no such epidemic occurred in the next G-150 period around 1700. The likely reason is that this time C was higher, contributing to immunity.

But the pattern here is very different from other indications of C, in that the early stages of declining C saw a marked *decrease* in infectious disease, well before the invention of antibiotics. To take one example, tuberculosis was a major killer in Western countries in the eighteenth and early nineteenth centuries, causing almost 25% of all deaths.[345] But between 1851 and 1935, tuberculosis mortality in England and Wales dropped from more than 300 per 100,000 people to fewer than 100, with a similar decline in the virulence of other infectious diseases.

Less known but quite striking is that TB declined *before* any effective medical treatments were in place.[346] This is not because it was eliminated like smallpox or polio, since even today an estimated one third of the world's population has been infected by TB.[347] Nor can it be fully explained by better public health measures, though these have been effective against diseases such as cholera. The invention of antibiotics has certainly reduced mortality, but in historical terms they have had far less impact.

The explanation for this conundrum can be found in the link between disease and stress, since stressed people are more susceptible to disease. Thus, people should be most resistant to disease when C is at a high point and stress at a low point. The theory of the civilization cycle predicts exactly when this should be. We have seen earlier that in the late nineteenth and early twentieth century there was a rapid fall in V and stress but only a minor fall in C. It was only from the 1960s onwards that the fall in stress slowed and that of C started to accelerate. It follows from this that the optimum time for high C and low stress should be around 1960, as indicated in Fig. 16.12 below. This is when resistance to disease should be greatest.

The opposite phase of the civilization cycle occurs when C is lowest relative to stress and we should expect least resistance to disease. As indicated in Fig. 16.12, this can be placed in the late fourteenth and early fifteenth centuries, the time of the Black Death.

Fig. 16.12. The civilization cycle continued into the twenty-first century. From the 1960s the decline of V and stress slowed and that of C accelerated. This marks the time when the balance of high C and low stress (C minus stress) is at a maximum. Any aspect of society which is promoted by *both* high C and low stress should peak at this time.

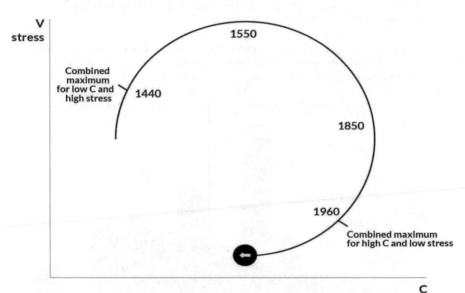

Thus it is, fifty years after the optimum point of the 1960s, that infectious disease has begun to rise. Since the 1980s there has been a resurgence of disease in many countries—not only old killers such as tuberculosis and cholera but new ones including Legionnaire's disease, HIV, hepatitis, SARS and avian influenza.[348] As yet this has been only a minor statistical change, nothing like a return to the early nineteenth century, but an ominous sign for the future as C continues to decline. An example of what to expect is the great plague of 165–180 AD that devastated the Roman Empire, and in the even greater one of Justinian's reign in the sixth century AD.

Economic Decline

Another factor that follows the same pattern is economic growth and decline. We have seen that high levels of C and especially infant C are associated with economic growth, and for machine skills and engineering this is almost certainly the strongest factor. But success in a technological society also depends on flexibility of mind, and such flexibility is inhibited by stress, especially as expressed in the punishment of older children (child V). Thus it is that the fastest growth rates in the twentieth century occurred not early in the century when C was highest, but between 1940 and 1970 when stress levels had dropped dramatically but before any major fall in C took place (Fig. 16.13).

Fig. 16.13. U.S. Growth Rates 1900–2013—average annual growth in real GDP per capita. The most rapid economic growth was between 1940 and 1970, bracketing the year 1960 when C was highest relative to Stress.[349]

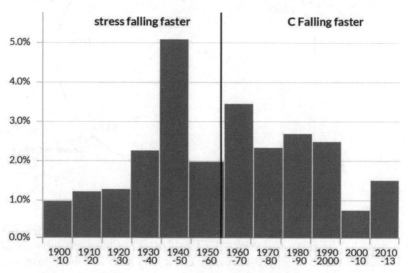

This mid-century prosperity resulting from falling stress masked the impact of declining C, but the economic effects are already being felt. Europe and Japan are stagnant, and only the US enjoys some growth - though nothing like that of the recent past.

Growing gap between rich and poor

Still another variable that follows this pattern is equality. Chapter six showed that as the economic skills of ordinary people increased from the Middle Ages, so the premium for skill and capital declined. Thus the gap between skilled and unskilled wages narrowed, and rents fell relative to land values.

This trend reached a peak around the 1960s, a time of low unemployment and relatively equal income distribution. But as C fell from the 1960s onward, certain groups were far more affected and their economic value declined. Among lower income groups, divorce rates are higher and married people less happy.[350] Unmarried and divorced parents are especially likely to exercise minimal discipline. As quoted in one community study of a Pennsylvania town:

> Then you hear why the discipline was only minimal—"Well, you know, I talked to them and they said this, that, and the other, and I figure "Maybe he's right" ... You want to be the cool parent, the friend parent, the great parent that the kid does whatever he wants, however he wants, dresses great.[351]

The essence of high C is hard work and the willingness to invest for the future. High C people are also more likely to be religious and have stable family lives. In the twenty-first century all these characteristics of middle-class families are less and less evident in the poor, who have increasingly come to depend on government handouts.

Thus it is that equality in Western countries reached a maximum in the years around 1960, after which inequality began to rise (see Fig. 16.14 below).

Fig. 16.14. Share of total market income, including capital gains, going to the top 10%.[352] A fall in Stress permitted greater equality in the first century after the peak of C. After that, rapidly declining C made much of the population less capable and so inequality increased. Increased inequality was also characteristic of the late Roman Republic.

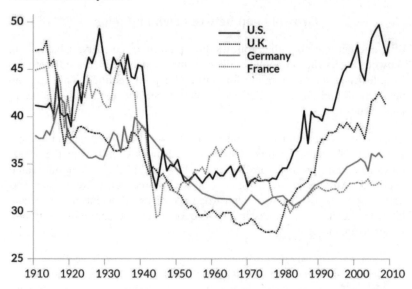

This is actually a return to an older pattern. In 1928 the richest 1% of Americans earned almost 20% of national income. Their share fell to around 7% by the 1970s but has risen to 16% in the last decade. The US has experienced the greatest rise in the super-wealthy but even relatively egalitarian countries such as Germany, Denmark and Sweden have experienced a rise in the income gap. In these countries the richest 10% out earned the poorest 10% by 6 to 1, compared to around 5 to 1 in the 1980s. By the same measure the wage gap in Italy, South Korea and Japan is currently 10 to 1, and in the US and Israel 14 to 1.[353]

These same trends were evident in the late Roman Republic, with the agrarian middle class disappearing and a huge gap opening between wealthy plutocrats and an urban mob living on corn doles. Western society has not yet degenerated to this extent, but all evidence suggests that it will.

The End of the West?

Biohistory supports scholars such as Gregory Clark in proposing that the rise of the West was driven not by economic or political forces but by a change in *temperament*. Between the twelfth and nineteenth centuries people became harder working, more willing to invest in education and skills, thriftier, more skilled at business, better with machinery, more innovative, more disciplined, and more supportive of large and stable political units. In other words, they showed all the attributes of rising C. The result was a huge surge of economic growth, starting with the Industrial Revolution in England, which transformed the world.

Over the past 150 years and especially in the last 50, the tide has turned and C has begun to fall. In the early stages of decline the economic impact was reduced by an even larger fall in stress and child V, helping people to become more flexible and open in their attitudes. Thus economic growth continued for a time and even accelerated, aided by the immigration of people from higher C countries such as India and China.

But in recent times economic growth has all but halted in most Western countries. The Japanese economy, without the benefit of large-scale immigration, has been stagnant since the 1990s. As C continues to fall and people become less and less capable economically, growth must turn into stagnation and then decline. Western nations will become weaker, with shrinking populations and growing poverty. Governments must inflate the currency to pay their debts, resulting in hyper-inflation.

Technology cannot save us because technology creates the wealth that is destroying us, far more than in the case of Rome. In the late Republic mobs rioted when the high price of bread left them hungry. In the modern West one of the greatest problems of the poor is *obesity*. So not only must Western economies collapse, but the collapse is likely to be much faster than was the case for Rome.

There will also be political changes. Democratic government depends on loyalty to impersonal institutions such as parliaments, the rule of law and electoral systems. All of these in turn rest on infant C, which is plunging fast. There has already been a decline of confidence in the political system, as measured by voting activity and membership of political parties. In America, the decline in the rule of law is expressed in the practice of reinterpreting the Constitution to fit "modern" values, such as those related to capital punishment or abortion. If the Constitution no longer follows

common-sense understanding or the obvious meaning of those who wrote
it, then it can be interpreted away by any President who controls the
Supreme Court. At that stage, republican government may remain in form
but the reality has died.

In the case of Europe, the end of democracy is as likely to come from
without as within. The nation and city states of ancient China, India and
Greece merged into cosmopolitan empires before their collapse. European
nations are already losing their sense of national identity and pride in their
own culture, and their shrinking and aging populations are less and less
keen on military service. As economies decline and democracy flounders,
a more vigorous neighbor such as Russia or some Middle Eastern power
must move in and take over. This last option is especially likely because
by then much of the population will be of Middle Eastern or African
origin.

It is perhaps some comfort to know that Europe will not suffer the anarchy
that followed the fall of Rome. The Muslim peoples of the Middle East
are numerous and growing, and continued mass immigration will have
profound demographic and cultural effects. The end of Europe may be
more like that of Byzantium in the fifteenth century, with a smooth
transition to a new Muslim Caliphate.

North America might follow a similar path, or perhaps develop some
rigidly conservative Christian version. Meanwhile the wealthy urban
populations of East Asia and elsewhere will decline and diminish, falling
prey to autocrats or collapsing into anarchy, the survivors drifting back to
subsistence farming in the struggle to stay alive. Western civilization, in
Europe and America and throughout the world, will collapse.

Eventually, as poverty and the revival of traditional values do their work,
V and C will revive. But this will be a very different world, with no more
brilliant low-S civilizations to create industrial wealth. To our
descendants, struggling for survival in a harsh world of famine and
disease, the wonders of our time will be no more than a distant memory.

Testing

Widespread physiological testing of Western populations could measure
how fast C is declining. Steadily lower levels of C would be expected in
successive age cohorts.

CHAPTER SEVENTEEN

THE FUTURE

It is a common belief that an individual from any background can succeed economically, given the right education. This means that societies fail to flourish only because their institutions and governments hold them back. If the government and institutions can be changed along western lines, the society will become wealthy and democratic. In other words, aside from levels of knowledge and cultural habits, people in different societies are basically the same.

Biohistory takes issue with this view. It proposes that human behavior and attitudes are largely determined by physiology. People in different societies and in different periods of history are different at the emotional and behavioral levels, and economic and political systems reflect these differences. Temperament is set in childhood and even to some extent at birth, and can only be changed in limited ways by later experience. Western forms of government and economic systems will not work in societies where the people are unsuited to them by temperament. And people without the appropriate temperament cannot function well in Western societies, regardless of education. Nor can they become wealthy.

It is vital to emphasize once more that this is not a matter of genes. The key differences between societies are not genetic but *epigenetic*.

The implications of biohistory

Chapter sixteen showed that the future of the West looks bleak. Levels of C and V, following the course of the civilization cycle, are in long-term decline. Such a decline, over the next century or two, must bring about the end of Western civilization. After centuries of world domination, the West has had its time. The future belongs to the vigorous, high-V cultures of the world, particularly Islam.

Conventional measures can do nothing to halt the decline. Financial incentives cannot raise the birth rate when people do not *want* more children. Education cannot teach people what they do not have the temperament to learn, such as habits of innovation and hard work. Economic stimulus packages cannot make people more productive. The slickest political campaigns cannot win support for governments from people who are no longer psychologically equipped to provide it. Pro-market economic policies can only unleash productive forces when such forces are available to be unleashed, and declining C populations have less and less taste for such policies in any case.

A nation's potential for economic growth is limited by the epigenetically-set temperament of the population. Chinese people have relatively high C and so are capable of rapid industrialization, held back by Maoist policies until the reforms of Deng Xiaoping unleashed their potential. By contrast, sub-Saharan Africa remains in poverty under a variety of socialist and capitalist regimes, and despite uncounted billions in aid. Government policy can thus be seen mainly as a *negative*. It can limit growth below the economic potential of the people, but cannot raise it beyond that level.

Likewise, a country's form of government depends primarily on the temperament of the population. Therefore, attempts to impose liberal democracy on countries such as Afghanistan and Iraq are doomed to fail, as the underlying temperament of the people cannot support such a government. When people are psychologically primed to accept brutal authority, removing such authority leads only to anarchy.

Major wars are driven less by social and political concerns than by the underlying temperament of the population as a whole, especially men in their early twenties. Economic booms and recessions can be understood in similar ways. This implies that physiological measurements can predict both wars and recessions.

Through understanding biohistory we gain new insight into the way religions function, and why they are so successful and widespread. Far from being an outdated relic, biohistory shows religious practice to be the key driver of the high C temperament. In particular, Christianity can be seen as responsible for the rise of the modern West, democracy and the scientific revolution. The decline of Christianity, and the moral principles it espouses, are thus a grim portent for the future.

Can we stabilize C?

The decline of Western civilization is happening in the same way and for the same reasons as has been the case for wealthy urban societies throughout history.

One of the major lessons of biohistory is that governments have less power to direct affairs than they believe. In reality, the future courses of our societies are determined by factors that we are only just beginning to understand. But now that such an understanding has begun, might it be possible to change the course of history? Now that have gained a knowledge of C and V, how they work, and how they can be promoted or undermined, could we manipulate them directly? Might we be able to slow the decline, or even halt or reverse it?

There is no simple answer to that question.

Part of the problem is that biohistory as a theory is still in its early stages of development. Though the basic mechanisms are fairly well understood, there is a great need for further scientific research. Another problem is that the measures needed to halt the decline of C may simply be unacceptable.

Diet, behavior and traditional culture

The most obvious way to stabilize C is through the adoption of C-promoting diet and cultural technologies. As we have seen, there are several forms of human behavior that promote C, from fasting to repetitive rituals to sexual abstinence. The adoption of more ascetic lifestyles, dietary restriction, regular performance of rituals, limits on sexual behavior, and strict regulation of daily life are all things which could, in principle, stabilize or raise a society's level of C. As we have seen, it is these C-promoting behaviors and restrictions that have supported civilization throughout history.

But promoting such behavior is extremely difficult, especially in a wealthy urban society. Even governments committed to conservative social values, such as Emperor Augustus in ancient Rome or Vladimir Putin in Russia, can do little to change personal behavior. But in any case, the dominant culture of the modern West reflects the rapid decline of C and is increasingly hostile to traditional values. Government schools, universities and the mass media tend to encourage behavior that undermines C, and governments increasingly penalize higher C behavior (such as by prohibiting strict discipline in schools). People with traditional values are an increasingly small and embattled minority.

So, if restoring traditional values is out of the question, could there be another way to stabilize C? A physiological one, perhaps?

C promotion through biochemistry

Preliminary studies on rats have shown promising results for the development of C-promoting supplements, as shown by improvements in maternal behavior and reduced consumption of alcohol. The potential for an effective C-promoter in humans is highly speculative, but the possibility opens up a number of interesting questions.

Would it be possible for someone to maintain their level of C solely with C-promoting supplements, or might it only work in conjunction with participation in a C-promoting culture? The principle of the effect feedback cycle suggests that a combination of the two may be the most effective.

Historically, high C has only been possible with high levels of V and stress, which create the rigidity that allows high C cultures to exist. Unfortunately, high stress also promotes rigid thought patterns, acceptance of authoritarian government, patriarchy and the punishment of children. But C-promoting supplements should have none of these drawbacks.

If an individual chose to take C-promoting supplements, what might their effects be? For a start, they would be more positive about family and more likely to want children. They would tend to avoid high-stimulus environments and thus experience lower anxiety. Addictive drugs, alcohol, tobacco and high calorie food might become less attractive. They would be less likely to indulge in criminal or delinquent behavior. They might be more resistant to disease and have fewer problems with infertility.[354] They would be better educated and more effective workers, also more likely to start and run successful businesses. Combining high C with a low level of stress-based rigidity, they could be more creative. They might also be more religious. The exact range of effects would depend heavily on the age at which the C-promoter was applied.

The most likely initial applications for a C-promoting supplement would be the treatment of conditions such as obesity, anxiety and drug addiction. Other uses would be more controversial.

On the negative side, C-promoters might make people less gregarious, more inhibited and should reduce libido in adults. Given this, it is likely that only a minority of people would be interested. It need hardly be said that any attempt to impose C-promoters on people who did not want them would be abhorrent.

Stabilizing V

Although higher stress would be undesirable, individuals might benefit from a slightly higher level of V. V reflects a "toughening" of the stress response which allows individuals to react more effectively to challenge. Increasing V while maintaining a low level of chronic stress (such as through C-promoters) could be an especially effective treatment for anxiety and depression. A higher level of V might also make C-promoters more effective.

However, increasing V excessively would be dangerous. In the past, high levels of V allowed our ancestors to organize and defend themselves against outsiders. Today, such high levels of aggression would be severely counterproductive. A particular danger is that an autocratic regime might choose to raise the level of V in the population as a whole and thus cause an outbreak of war. As suggested in chapter nine, a surge of V was responsible for such disasters as two World Wars and the Chinese Cultural Revolution.

This is a particular concern since, as will be clear to anyone with expertise in the field, raising the level of V would be relatively easy.

Conclusion

This book has outlined a theory which explains historical and social change as a result of physiological changes that affect temperament and behavior. We have explored the nature and some of the potential reasons for organized religion, traditional cultural practices such as fasting, rituals and patriarchy. These cultural technologies have helped induce physiological states, which we have termed C and V, which are necessary to build civilization.

C is a physiological system associated with behaviors that adapt animals to food-limited environments. These include monogamy, territoriality, intense care of young, and reduced sexual activity.

V is a physiological system associated with stress responses, and related behaviors such as aggression, that help animals survive in dangerous environments.

High C in one generation has a profound effect on the next, determining its attitudes towards law, the economy and political and religious beliefs. The transmission of C can be through parental behavior, influences within the womb, and possibly inherited epigenetic changes.

The rise and fall of civilizations follows the pattern of the civilization cycle, which represents the interaction of C and V over hundreds of years. The rise and fall of Ancient Rome comprised one civilization cycle, the rise and decline of the West another. Within these cycles are population booms and troughs associated with lemming cycles and recession cycles, which have a range of social and political effects.

In the context of the current cycle, our civilization is in decline due to rapidly falling levels of C and V. The immense momentum of these trends means that no current government programs can significantly affect the outcome. A society's temperament determines its course, not the mechanisms of economics or social relations.

However, now that we understand—or rather, are beginning to understand—the underlying reasons for the cycle, it may be possible to devise ways of diverting, altering or halting it, and preserving our technological civilization in its current form. The most likely way for this to happen is for a minority segment of the population to reinforce traditional values with some form of C-promoting supplement. Over time, their economic success and higher birth rate would influence the wider society, especially given its ongoing decline.

While the notion of cyclical patterns in the histories of civilizations is not entirely new, the biohistorical civilization cycle model is unique in that it offers a physiological framework that links history to the science of human nature. In this book we have examined and uncovered the reasons why these cycles exist, tracked them throughout human history, and made clear the steep downward slope that now lies before us. No conventional social or economic policies can prevent the eventual collapse of Western civilization.

But that does not make it inevitable. Assuming the theory is broadly correct, a dedicated program of scientific research would bring a much better understanding of the mechanisms at work. Given that, we may develop ways to hold off the collapse … if we choose to use them.

Afterword

As mentioned earlier, this book is a simplified introduction to biohistory. Students interested in a more detailed discussion and the full evidence are invited to read my book *Biohistory*, also published by Cambridge Scholars Publishing.

Anyone interested in furthering biohistory, especially biological scientists wishing to collaborate on the research program, is invited to contact the author through www.biohistory.org. Relevant research will be promoted by our upcoming biohistory journal, to be published as an open access journal through the biohistory web site.

NOTES

[1] Jim Penman, *Biohistory* (Newcastle: Cambridge Scholars Publishing, 2015).

[2] *The Economist*, "Briefing: the Future of Jobs," January 18, 2014.

[3] G. Clark, *A Farewell to Alms: A Brief Economic History of the World* (Princeton, NJ: Princeton University Press, 2007), Chapter 9.

[4] War in Iraq: Not a Humanitarian Intervention, Human Rights Watch, January 26, 2004.

[5] H. Ammar, *Growing up in an Egyptian Village* (New York: Octagon, 1966), 80–81.

[6] G. Lawton, "I believe: Your personal guidebook to reality", *New Scientist*, 4 April 2015

[7] Arthur Jensen, *The Bell Curve: Intelligence and Class Structure in American Life* (New York: Free Press, 2010)

[8] R, Bowden, T.S. MacFie, S. Myers, G. Hellenthal, E. Nerrienet, R. E. Bontrop, C. Freeman, P. Donnelly & N. Mundy, "Genomic Tools for Evolution and Conservation in the Chimpanzee: *Pan troglodytes ellioti* Is a Genetically Distinct Population," *PLOS Genetics* March 1, 2012.

[9] For a fuller account of the evidence, including a systematic cross-cultural survey of 67 societies, see J. Penman, *Biohistory* (Newcastle: Cambridge Scholars Publishing, 2015). The survey itself is detailed at www.biohistory.org. Other studies in this field include: J. D. Unwin, *Sex and Culture* (Oxford: Oxford University Press, 1934), 13–14, 27–9, 315–7, 321; W. N. Stephens, *The Family in Cross-cultural Perspective* (New York, Holt, Rinehart & Winston, 1963), 256–258; F. B. Aberle, "Culture and Socialization," in *Psychological Anthropology*, edited by F. L. K. Hsu, 386 (Homewood, Illinois: Dorsey Press, 1961); R. Barry, I. Child & M. K. Bacon, "The Relation of Child training to the Subsistence Economy," *American Anthropologist* 61 (1959): 56–60.

[10] R. K. Beardsley, J. W. Hall & R. E. Ward, *Village Japan* (Chicago: The University of Chicago Press, 1959).

[11] Ibid., 312–318.

[12] Ibid., 332.

[13] W. J. Smole, *The Yanomamo Indians: A Cultural Geography* (Austin: University of Texas Press, 1976), 73–74.

[14] This chapter contains only general references. For sources and a full discussion of the evidence see Penman, *Biohistory*.

[15] D. L. Cheney & R. M. Seyfarth, *Baboon Metaphysics: The Evolution of a Social Mind* (Chicago: University of Chicago Press, 2007).

[16] S. C. Albert, J. C. Buchan & J. Altmann, "Sexual Selection in Wild Baboons: From Mating Opportunities to Paternity Success," *Animal Behaviour* 72 (5) (2006): 1177–96.

[17] Cheney & Seyfarth, *Baboon Metaphysics*.

[18] U. Reichhard, *Social Monogamy in Gibbons: The Male Perspective. Monogamy: Mating Strategies and Partnerships in Birds, Humans and other Mammals*

(Cambridge: Cambridge University Press, 2003), 192; N. L. Uhde & V. Sommer, "Antipredator Behavior in Gibbons (Hylobates lar, Khao Yai/Thailand)," in *Eat or be Eaten: Predator Sensitive Foraging among Primates*, edited by L. E. Miller (Cambridge: Cambridge University Press, 2002).

[19] Uhde & Sommer, "Antipredator Behavior in Gibbons"; W. Y. Brockelman, U. Reichhard, U. Treesucon & J. J. Raemaekers, "Dispersal, Pair Formation and Social Structure in Gibbons (Hylobates lar)," *Behavioral Ecology and Sociobiology* 42 (5) (1998): 329–39; T. Geissmann, "Reassessment of age of sexual maturity in gibbons (Hylobates spp.)," *American Journal of Primatology* 23 (1) (1991): 11–22.

[20] R. A. Palombit, "Longitudinal Patterns of Reproduction in Wild Female Siamang (Hylobates syndactylus) and White-Handed Gibbons (Hylobates lar)," *International Journal of Primatology* 16 (5) (1995): 739–60.

[21] Detailed references for this chapter can be found in J. Penman, *Biohistory* (Newcastle: Cambridge Scholars, 2015).

[22] C. Turnbull, *The Forest People* (London: The Reprint Society, 1963), 38–9.

[23] J. S. Ruff., A. L. Suchy, S. A. Hugentobler, M. M. Sosa, B. L. Schwartz, L. C. Morrison, S. H. Gient, M. K. Shigenaga & W. K. Potts, "Human-Relevant Levels of Added Sugar Consumption Increase Female Mortality and Lower Male Fitness in Mice," *Nature Communications* 4 (2013): Article 2245.

[24] Penman, *Biohistory*.

[25] J. M. Dabbs, *Heroes, Rogues and Lovers: Testosterone and Behaviour* (New York: McGraw Hill, 2000): 150–151.

[26] J. M. Dabbs, D. de La Rue & P. M. Williams, "Testosterone and Occupational Choice: Actors, Ministers, and Other Men," *Journal of Personality and Social Psychology* 59 (6) (1990): 1261–5.

[27] E. A. Levay, A. G. Paolini, A. Govic, A. Hazi, J. Penman & S. Kent, "Anxiety-like Behaviour in Adult Rats Perinatally Exposed to Maternal Calorie Restriction," *Behavioural Brain Research* 191 (2) (2008): 164–72.

[28] C. M. Turnbull, *Wayward Servants: The Two Worlds of the African Pygmies* (Westport: Greenwood Press, 1976), 39–40.

[29] R. T. Ridley, *History of Rome, a Documented Analysis* (Rome: l'Erma di Bretschneider, 1987), 215–225, 236.

[30] G. Clark, *A Farewell to Alms: A Brief Economic History of the World* (Princeton, NJ: Princeton Diamond University Press, 2007).

[31] E. Westermarck, "The Principles of Fasting," *Folklore* 18 (4) (1907): 391–422.

[32] F. Azizi, "Medical Aspects of Islamic Fasting," *Medical Journal of the Islamic Republic of Iran* 10 (3) (1996): 241–246.

[33] *The Hindu*, Online Edition, August 20, 2007

[34] For a full list of sources and a more complete argument, see J. Penman, *Biohistory* (Newcastle: Cambridge Scholars Publishing, 2015).

[35] J. M. Dabbs, *Rogues and Lovers: Testosterone and Behavior* (New York: McGraw Hill, 2000), 102.

[36] J. M. Graham & C. Desjardins, "Classical Conditioning Induction of Luteinizing Hormone and Testosterone Secretion in Anticipation of Sexual Activity," *Science* 210 (4473) (1980): 1039–41; S. G. Stoléru, A. Enaji, A. Cournot & A. Spira,

"Pulsatile Secretion and Testosterone Blood Levels are Influenced by Sexual Arousal in Human Males," *Psychoneuroendocrinology* 18 (3) (1993): 205–18.

[37] A. C. Kinsey, W. B. Pomeroy & C. E. Martin, *Sexual Behavior in the Human Male* (Philadelphia: W. B. Saunders Company, 1948): 466–467; M. Zborowski & E. Herzog, *Life is with People* (New York: International University Press, 1952); Glasenapp, H. v., *Jainism: An Indian Religion of Salvation* (Delhi: Motilal Banarsidass, 1999), 228–31.

[38] D. Wu & A. C. Gore "Sexual Experience Changes Sex Hormones but not Hypothalamic Steroid Hormone Receptor Expression in Young and Middle-aged Male Rats," *Hormones and Behavior* 56 (3) (2009): 299–308.

[39] Kinsey et al., *Sexual Behavior in the Human Male*, 342, 420–5.

[40] Further references in Penman, *Biohistory*.

[41] G. Cochrane & H. Harpending *The 10,000 Year Explosion: How Civilization Accelerated Human Evolution* (New York: Basic Books, 2009); Nicholas Wade, *A Troublesome Inheritance: Genes, Race and Human History* (New York: Penguin, 2014).

[42] S. J. Solnick & David Hemenway, "The 'Twinkie Defense': the Relationship Between Carbonated Non-diet Soft Drinks and Violence Perpetration among Boston High School Students," *Injury Prevention* 040117 (2011); S. J. Solnick & David Hemenway, "Soft drinks, aggression and suicidal behavior in US high school students," *Int J Inj Contr Saf Promot. Epug* (July 8, 2013), A. Siferlin, "Soda Contributes to Behavior Problems Among Young Children" *TIME*, August 16, 2013 http://healthland.time.com/2013/08/16/soda-contributes-to-behavior-problems-among-young-children/?iid=hl-main-lead (accessed September 4, 2014)

[43] C. Woodham-Smith, *The Reason Why* (London: Constable, 1953), 223–262.

[44] N. M. Chagnon, *The Yanomamo* (Fort Worth: Harcourt Brace College Publishers, 1997); W. J. Smole, *The Yanomamo Indians: A Cultural Geography* (Austin: University of Texas Press, 1976), 73–74.

[45] Ibid., 205.

[46] M. Mead, *Sex and Temperament in Three Primitive Societies* (New York: Harper Perennial, 2001), 232

[47] D. E. Brown, *Human Universals* (New York: McGraw Hill, 1991); Education Portal, "Tchambuli Tribe: Culture, Gender Roles & Lesson," http://education-portal.com/academy/lesson/tchambuli-tribe-culture-gender-roles-lesson.html# lesson (accessed September 3, 2014).

[48] D. L. Cheney & R. M. Seyfarth, *Baboon Metaphysics: The Evolution of a Social Mind* (Chicago: University of Chicago Press, 2007).

[49] S. Stueckle & D. Zinner, "To Follow or Not to Follow: Decision Making and Leadership During the Morning departure in Chacma Baboons," *Animal Behavior* 75 (6) (2008): 1995–2004; A. J. King, C. M. S. Douglas, E. Huchard, N. J. B. Isaac & G. Cowlishaw, "Dominance and Affiliation Mediate Despotism in a Social Primate," *Current Biology* 18 (23) (2008): 1833–8.

[50] C. Sueur, "Group Decision-Making in Chacma Baboons: Leadership, Order and Communication during Movement," *BMC Ecology* 11 (2011): 26.

[51] S. Zuckerman, *Social Lives of Monkeys and Apes* (London: Kegan Paul, 1932).

[52] Cheney & Seyfarth, *Baboon Metaphysics*, 42.

[53] A. L. Engh, J. C. Beehner, T. J. Bergman, P. L. Whitten, R. R. Hoffmeier, R. M. Seymarth & D. L. Cheney, "Behavioral and Hormonal Responses to Predation in Female Chacma Baboons (Papio hamadryas ursinus)," *Proceedings of the Royal Society of London—Series B: Biological Sciences* 273 (1587) (2006): 707–12; Cheney & Seyfarth, *Baboon Metaphysics*, 57.

[54] S. H. Dhabhar & B. S. McEwen, "Enhancing Versus Suppressive Effects of Stress Hormones on Skin Immune Function," *PNAS* 96 (3) (1999): 1059–1064. A more extensive discussion on stress effects can be found in Penman, *Biohistory* (Newcastle: Cambridge Scholars Publishing, 2015).

[55] B. S. McEwen, "Central Effects of Stress Hormones in Health and Disease: Understanding the Protective and Damaging Effects of Stress and Stress Mediators," *European Journal of Pharmacology* 583 (2–3) (2008): 174–185.

[56] M. J. Weiss, P. A. Goodman, B. G. Losito, S. Corrigan. J. M. Charry & W. H. Bailey, "Behavioural Depression produced by an Uncontrollable Stressor: Relationship to Norepinephrine, Dopamine, and Serotonin Levels in Various Regions of Rat Brain," *Brain Research Reviews* 3 (2) (1981): 167–205; Sunanda, B. S. Rao & T. R. Raju, "Restraint Stress-Induced Alterations in the Levels of Biogenic Amines, Amino Acids, and AChE Activity in the Hippocampus," *Neurochemical Research* 25 (12): 1547–1552; B. S. McEwen, "Central Effects of Stress Hormones in Health and Disease: Understanding the Protective and Damaging Effects of Stress and Stress Mediators," *European Journal of Pharmacology* 583 (2–3) (2008): 174–185.

[57] K. A. Roth, I. M. Mefford & J. D. Barchas, "Epinephrine, Norepinephrine, Dopamine and Serotonin: Differential Effects of Acute and Chronic Stress on Regional Brain Amines," *Brain Research* 239 (2) (1982): 417–424.

[58] M. Tichomirowa, M. Keck, H. J. Schneider, M. Paez-Pereda, U. Renner, F. Holsboer & G. Stalla, "Endocrine Disturbances in Depression," *Journal Of Endocrinological Investigation* 28 (1) (2005): 89–99.

[59] A. Adell, C. Garcia-Marquez, A. Armario & E. Gelpi, "Chronic Stress Increases Serotonin and Noradrenaline in Rat Brain and Sensitizes their Responses to a Further Acute Stress," *Journal Of Neurochemistry* 50 (6) (1988): 1678–1681

[60] P. H. Wirtz, U. Ehlert, M. U. Kottwitz, R. La Marca & N. K. Semmer, "Occupational Role Stress is Associated With Higher Cortisol Reactivity to Acute Stress," *Journal of Occupational Health Psychology* 18 (2) (2013): 121–131.

[61] A. Adell, C. Garcia-Marquez, A. Armario & E. Gelpi, "Chronic Stress Increases Serotonin and Noradrenaline in Rat Brain and Sensitizes their Responses to a Further Acute Stress," *Journal Of Neurochemistry* 50 (6) (1988); 1678–1681; S. Jordan, G. Kramer, P. Zukas & F. Petty, "Previous Stress Increases in Vivo Biogenic Amine Response to Swim Stress," *Neurochemical Research* 19 (12) (1994): 1521–1525.

[62] J. B. Silk, D. Rendall, D. L. Cheney & R. M. Seyfarth, "Natal Attraction in Adult Female Baboons (Papio cynocephalus ursinus) in the Moremi Reserve, Botswana," *Ethology* 109 (8) (2003): 627–44.

[63] L. T. Nash, "The Development of the Mother-Infant Relationship in Wild Baboons (Papio anubis)," *Animal Behavior* 26 (1978): 746–759.

[64] J. P. Henry & P. M. Stephens, *Stress, Health and the Social Environment* (New York: Springer-Verlag, 1977), 118–41.

[65] H. F. Harlow & M. K. Harlow, "Maternal Behavior of Rhesus Monkeys Deprived of Mothering and Peer Associations in Infancy," *Proceedings of the American Philosophical Society* 110 (1) (1966): 58–66.

[66] M. Mead, *Sex and Temperament in Three Primitive Societies.*

[67] For more information, including how lower adult anxiety increases aggression, see Penman, *Biohistory*, chapter four.

[68] Chagnon, *The Yanomamo*, 126.

[69] Mead, *Sex and Temperament in Three Primitive Societies*, 237–238

[70] Further references in Penman, *Biohistory*, chapter four.

[71] Plutarch, *The Life of Lycurgus,* http://penelope.uchicago.edu/Thayer/E/Roman/Texts/Plutarch/Lives/Lycurgus*.html (downloaded September 9, 2014)

[72] J. Richards, *Visions of Yesterday* (London: Routledge and Kegan Paul, 1973), 40.

[73] R. Graves, *Goodbye to All That* (London: Penguin, 1958), 156.

[74] J. A. Smith, "Evaluation of Cortisol and DHEA as Biomarkers for Stress." Paper 626, (2008), http://scholarship.shu.edu/dissertations/1045/ (accessed September 3, 2014).

[75] H. Barry, I. L. Child & M. K. Bacon, "Relation of Child Training to Subsistence Economy," *American Anthropologist* 61(1) (1959): 51–63. Details of the author's cross-cultural survey can be seen in J. Penman, *Biohistory* (Newcastle: Cambridge Scholars Publishing, 2015).

[76] M. Mead, *Sex and Temperament in Three Primitive Societies* (New York: Harper Perennial, 2001); M. Mead, *Growing up in New Guinea* (New York: Perennial Classics, 2001).

[77] M. Toledo-Rodriguez & C. Sandi, "Stress before Puberty Exerts a Sex- and Age-related Impact on Auditory and Contextual Fear Conditioning in the Rat." *Neural Plasticity* (2007) (article ID 71203:1–12); G. E. Hodes & T. J. Shors, "Distinctive Stress Effects on Learning During Puberty," *Hormones and Behavior* 48 (2) (2005): 163–71.

[78] M. Matsumoto, H. Togashi, K. Konno, H. Koseki, R. Hirata, T. Izumi, T. Yamaguchi & M. Yoshioka, "Early Postnatal Stress Alters the Extinction of Context-Dependent Conditioned Fear in Adult Rats," *Pharmacology Biochemistry and Behavior* 89 (3) (2008): 247–52.

[79] S. Mansfield, *The Mormonizing of America: How the Mormon Religion became a Dominant Force in Politics, Entertainment and Pop Culture* (Brentwood, Tennessee: Worthy Publishing, 2012), Introduction.

[80] *The Hindu*, Online Edition, August 20, 2007.

[81] M. Gladwell, *Outliers* (New York, London: Little, Brown & Company, 2008).

[82] "How do we Know Leonardo was Gay?"

www.bnl.gov/bera/activities/globe/leonardo_da_vinci.htm (accessed September 3, 2014).

[83] A. Hughes, *Michelangelo* (London: Phaidon, 1997), 326.

[84] A. Storr, "Isaac Newton," *British Medical Journal (Clinical Research Edition)* 291(6511) (1985): 1779–1784.

[85] Michael White & John Gribbin, *Stephen Hawking: A Life in Science* (2nd ed.), (Washington DC: National Academies Press, 2002); T. Bridget, "The Impact of Assistive Equipment on Intimacy and Sexual Expression," *The British Journal of Occupational Therapy* 74 (9) (2011): 435–442.

[86] J. Gathorne-Hardy, *The Unnatural History of the Nanny* (New York: Dial Press, 1973), 263.

[87] N. Bakker, "The Meaning of Fear. Emotional Standards for Children in the Netherlands, 1850–1950: Was there a Western Transformation?" *Journal of Social History* 34 (2) (2000): 371–3.

[88] E. Kloek, "Early Modern Childhood in the Dutch Context," in *Beyond the Century of the Child: Cultural History and Developmental Psychology*, edited by W. Koops & M. Zuckerman, 53 (Philadelphia: University of Pennsylvania Press, 2012).

[89] A. H. Smith, *Village Life in China: A Study in Sociology* (New York: Fleming H. Revell, 1899), 237–238.

[90] M. Wolf, *Women and the Family in Rural Taiwan* (Stanford: Stanford University Press, 1972), 67–70.

[91] H. Ammar, *Growing up in an Egyptian Village* (New York: Octagon, 1966), 26.

[92] R. K. Beardsley, J. W. Hall & R. E. Ward, *Village Japan* (Chicago, The University of Chicago Press, 1959), 294.

[93] R. J. Smith & E. L. Wiswell, *The Women of Suye Mura* (Chicago: University of Chicago Press, 1982), 212–220; J. F. Embree, *Suye Mura: a Japanese Village* (Chicago: University of Chicago Press, 1939), 185.

[94] H. P. Varley, *Imperial restoration in Medieval Japan* (New York, Columbia University Press, 1971), 557–9; M. Takizawa, *The Penetration of the Money Economy in Japan* (New York: Amis, 1927), 34, 36; Smith (1959); L. Frédéric, *Daily Life in Japan at the Time of the Samurai* (New York: Praeger, 1972); J. W. Hall, *Government and Local Power in Japan: 500–1700* (Princeton: Princeton University Press, 1966); C. J. Dun, *Everyday Life in Traditional Japan* (London: Batesford, 1969), 20; W. Cole, *Kyoto in the Monoyama Period* (Norman, Oklahoma: University of Oklahoma Press, 1967), 74–75; G. Clark, *A Farewell to Alms: A Brief Economic History of the World* (Princeton, NJ: Princeton University Press, 2007), ch. 13.

[95] Beardsley et al., *Village Japan*, 294.

[96] Ammar, *Growing up in an Egyptian Village*, 137–138.

[97] Wolf, *Women and the Family in rural Taiwan*, 67–70.

[98] Full references in Penman, *Biohistory*, chapter five.

[99] Ammar, *Growing up in an Egyptian Village*, 80–81.

[100] Rat experiments suggest that stresses in the juvenile period, equivalent to late childhood in humans, caused an increase in conditioned fear that was harder to

extinguish in adults. Toledo-Rodriguez & Sandi, "Stress before Puberty Exerts a Sex- and Age-related Impact on Auditory and Contextual Fear Conditioning in the Rat." G. E. Hodes & T. J. Shors, "Distinctive Stress Effects on Learning During Puberty," *Hormones and Behavior* 48 (2) (2005): 163–71.

[101] Ammar, *Growing up in an Egyptian Village*, 79.

[102] Mead, *Growing up in New Guinea*, 22.

[103] Ibid., 27.

[104] Ibid., 38.

[105] M. Mead, *New Lives for Old* (New York: Perennial Classics, 2001), 170.

[106] Mead, *Growing up in New Guinea*, 154.

[107] K. Kais, "The Paliau Movement." Buai Digital Project (April 1998). http://www.pngbuai.com/100philosophy/paliau-movement/ (accessed September 3, 2014).

[108] Mead, *Growing up in New Guinea*, 358–359.

[109] See Ricardo Duchesne, *The Uniqueness of Western Civilization* (Leiden and Boston: Brill, 2012) for a summary of other approaches.

[110] Karras, Sexuality in Medieval Europe, Doing Unto Others (New York: Routledge, 1985), 20–22, 43, 44.

[111] P. N. Stearns, *Sexuality in World History* (Oxon, Canada: Routledge, 2009), 48.

[112] William Acton, *The Functions and Disorders of the Reproductive Organs in Childhood, Youth, Adult Age, and Advanced Life: Considered in Their Physiological, Social, and Moral Relations*, 3rd Edition (London: Churchill, 1862), 101.

[113] G. Clark, *A Farewell to Alms: A Brief Economic History of the World* (Princeton, NJ: Princeton University Press, 2007), loc. 815, 936, 961, 999, 1030.

[114] J. Hagnal, "European Marriage Patterns in Perspective," in *Population in History*, edited by D. V. Glass & D. E. C. Eversley (London: Edward Arnold, 1965), 120; E. A. Wrigley, "The Growth of Population in Eighteenth Century England: A Conundrum Solved," *Past and Present* 98 (1983): 131.

[115] T. H. Hollingsworth, "A Demographic Study of the British Ducal Families," in *Population in History*, edited by D. V. Glass & D. E. C. Eversley (London: Edward Arnold, 1965).

[116] R. Woods, *The Demography of Victorian England and Wales* (Cambridge: Cambridge University Press, 2000), 108; E. A. Wrigley, "Variation in Mean Age of First Marriage Among 10 English Parishes, 1551–1837," in *The Population History of England 1541–1871: A Reconstruction*, edited by R. S. Schofield (London: Edward Arnold, 1984); Data from 1850–1901 from E. A. Wrigley & R. S. Schofield, *The Population History of England 1541–1871* (London: Edward Arnold, 1981), 437, and averaged out between six figures ranging from 1851–1901 to fit the format of the previous data. Data comprises England only, minus Monmouth.

[117] J. Tosh, A Man's Place: Masculinity and the Middle-class Home in Victorian England (New Haven and London: Yale University Press, 1999), 13.

[118] F. R. H. DuBoulay *An Age of Ambition* (London: Nelson, 1970), 116; H. M. Smith, *Pre-reformation England* (London: MacMillan, 1963), 235; L. B. Wright, *Middle Class Culture in Elizabethan England* (London: Methuen, 1935), 203; L.

Stone, *The Family, Sex and Marriage in England: 1500–1800* (London: Weidenfeld & Nicholson, 1977); M. Girouard, *Life in the English Country House* (New Haven and London: Yale University Press, 1978), 219, 285, 298.

[119] quoted in Tosh, A Man's Place, 29.

[120] Ibid., 21, 23, 24, 29, 30, 32, 33, 38, 39.

[121] C. Z. Stearns, "'Lord Help Me Walk Humbly': Anger and Sadness in England and America, 1570–1750," in *Emotion and Social Change*, edited by C. Z. Sterns & P. N. Sterns, 51 (New York: Holmes and Meier, 1988).

[122] H. Tingsten, *Victoria and the Victorians* (London: George Allen & Unwin, 1972), 27.

[123] P. A. W. Collins, *From Manly Tear to the Stiff Upper Lip: The Victorians and Pathos* (Wellington: Victoria University Press, 1974), 17.

[124] Susanna Wesley, letter to her son, July 24, 1732, in Charles Wallace Jr (ed.), *Susanna Wesley: The Complete Writings* (New York: Oxford University Press, 1997), 369.

[125] C. Turnbull, *The Forest People* (London: Reprint Society, 1961), 92.

[126] E. Rodgers, *Discussion of Holidays in the later Middle Ages* (New York: Columbia University Press, 1940), 10–11; C. R. Cheney, "Rules for the Observance of Feast-days in Medieval England," *Bulletin of the Institute of Historical Research* 34 (90) (1961): 117–47.

[127] D. A. Reid, "The Decline of St Monday: 1766–1876," *Past and Present* 71 (1976): 76–101; E. P. Thompson, *The Making of the English Working Class* (London: Pelican, 1968), 321–322, 337–338, 345, 351.

[128] Based on 69-hour week: W. S. Woytinsky, Hours of labor. Encyclopedia of the Social Sciences, vol. III, (New York: Macmillan, 1935). A low estimate assumes a 45-week year, a high one assumes a 52-week year. Calculated from Bureau of Labor Statistics data, Office of Productivity and Technology.

[129] Clark, *A Farewell to Alms*, ch. 9.

[130] M. Higgs, Life in the Victorian and Edwardian Workhouse (Gloucestershire: Tempus Publishing, 2007), 14, 20; F. Driver, *Power and Pauperism* (Cambridge: Cambridge University Press, 2004), 59.

[131] Driver, *Power and Pauperism*, 59.

[132] B. Fagan, *How Climate made History, 1300–1850* (New York: Basic Books, 2000).

[133] R. C. Allen, *The British Industrial Revolution in a Global Perspective* (Cambridge: Cambridge University Press, 2009).

[134] B. Bunch & A. Hellemans, *The History of Science and Technology: A Browser's Guide to the Great Discoveries, Inventions, and the People who made them from the Dawn of Time to Today* (New York: Houghton Mifflin Company, 2004).

[135] C. Murray, *Human Accomplishment: The Pursuit of Excellence in the Arts and Sciences, 800 BC to 1950* (New York, NY: Harper Collins, 2003), 347; I. Asimov, *Asimov's Chronology of Science and Discovery* (New York: Harper Collins, 1994); B. L. Gary, "A New Timescale for Placing Human Events, Derivation of Per Capita Rate of Innovation, and a Speculation on the Timing of the Demise of

Humanity" (Unpublished Manuscript, 1993); P. A. Sorokin, *The Crisis of our Age: The Social and Cultural Outlook* (Boston: E.P. Dutton, 1942).

[136] J. Huebner, "A Possible Declining Trend for Worldwide Innovation," *Technological Forecasting and Social Change* 72 (8) (2005): 980–6.

[137] J. W. Hall, *Government and Local Power in Japan: 500–1700* (Princeton: Princeton University Press, 1966), 257, 291; E.O. Reischauer & J.K. Fairbank, *East Asia: the Great Tradition* (Boston, Houghton Mifflin, 1960, 628. T. C. Smith, *The Agrarian Origins of Modern Japan* (Stanford: Stanford University Press, 1959), 145–146.

[138] R. J. Smith, "Town and city in pre-modern Japan: Small families, small households, and residential instability," *Urban Anthropology*, edited by A. Southall, 180 (Oxford: Oxford University Press, 1973); H. Befu, "Origin of Large Households and Duolocal Residence in Central Japan," *American Anthropologist* 70 (2) (1968): 309–19.

[139] M. Ohta, "The Discovery of Childhood in Tokugawa Japan", *Wako University: Bulletin of the Faculty of Human Studies*, No.4, 2011

[140] Reischauer & Fairbank (1960), 484

[141] Ibid., 557–561, 626–8, 661.

[142] Ibid., 482–3.

[143] P. Das, "Shakespeare's Representation of Women in his Tragedies," *Prime University Journal* 6 (2) (2012); S. Doran, "Elizabeth I: Gender, Power and Politics," *History Today* 53 (5) (2003); S. Broomhall, *Women and the Book Trade in Sixteenth-Century France* (Aldershot: Ashgate Publishing, 2002).

[144] J. Tosh, A Man's Place: Masculinity and the Middle-class Home in Victorian England (New Haven and London: Yale University Press, 1999), 39.

[145] L. DeMause, "The Evolution of Childhood," in *The History of Childhood*, edited by L. DeMause, 132, 137 (New York: The Psychohistory Press, 1974).

[146] Brinsley, quoted in M. J. Wiener, *Men of Blood: Violence, Manliness and Criminal Justice in Victorian England* (Cambridge: Cambridge University Press, 2004).

[147] M. V. C. Alexander, *The Growth of English Education, 1348–1648. A Social and Cultural History* (London: The Pennsylvania State University Press, 1990), 199, 200.

[148] Ibid., 235.

[149] Wiener, *Men of Blood*, 53, 54.

[150] Ibid., 6.

[151] *Foxe's Book of Martyrs*, Chapter XVI. (Altermunster, Jazzybee Vergal Jurgen Beck, 2012)

[152] W. Holdsworth, *A History of English law* 3rd edn. (London: Methuen, 1945), 194; J. Lawrence, *A History of Capital Punishment: With Special Reference to Capital Punishment in Great Britain* (Port Washington: Kennikat Press, 1971), 408.

[153] L. Radzonowicz, *A History of Criminal Law and its Administration from 1750* (London: Stevens & Co., 1948), 139–142, 149–152.

[154] P. M. Rieder, *On the Purification of Women: Churching in Northern France, 1100–1500* (New York: Palgrave/Macmillan, 2006), 155–56.

[155] L. Stone, *The Crisis of the Aristocracy* (Oxford: Oxford University Press, 1965), 21, 34–5, 591–2.

[156] Stone, *The Crisis of the Aristocracy*, 35–6, 747–53; M. Girouard, Life in the English Country House (New Haven and London: Yale University Press, 1978), 85–6, 182–3, 231, 308; L. Stone, *The Family, Sex and Marriage in England 1500–1800* (London: Weidenfeld, 1977).

[157] Wiener, *Men of Blood*, 5; P. Carter, *Men and the Emergence of Polite Society, 1660–1800* (Harlow: Longman, 2000); A. Bryson, *From Courtesy to Civility: Changing Codes of Conduct in Early Modern England* (Oxford: Oxford University Press, 1998).

[158] J. A. Williamson, The Tudor Age (London: Longmans, Green & Co., 1953), 8–9.

[159] C. Bridenbaugh, *Vexed and Troubled Englishmen, 1590–1642: The Beginnings of the American People* (Oxford: Clarendon Press, 1968), 189.

[160] Stone, *The Family, Sex and Marriage in England: 1500-1800* (London, Weidenfeld & Nicolson, 1977) 93–96.

[161] Wiener, *Men of Blood*, 6.

[162] Ibid.

[163] G. B. Gibson, *Japan, A Short Cultural History* (London: The Crescent Press, 1931), 431.

[164] M. Ohta, "The Discovery of Childhood in Tokugawa Japan," *Wako University: Bulletin of the Faculty of Human Studies* 4 (2011).

[165] D. Chitty, *Do Lemmings Commit Suicide? Beautiful Hypotheses and Ugly Facts* (Oxford: Oxford University Press, 1996).

[166] C. H. D. Clarke, "Fluctuations in Populations," *Journal of Mammalogy* 30 (1) (1949): 21–5.

[167] L. B. Keith, *Wildlife's Ten-Year Cycle* (Madison: University of Wisconsin Press, 1963); P. L. Errington, *Muskrat Populations* (Ames: Iowa State University Press, 1963); J. J. Christian, "The Adreno-Pituitary System and Population Cycles in Animals," *Journal of Mammalogy* 31 (3) (1950): 247–59; J. Erb, N. C. Stenseth & M. S. Boyce "Geographic Variation in Population Cycles of Canadian Muskrats (Ondatra zibethicus)," *Canadian Journal of Zoology* 78 (6) (2000): 1009–16.

[168] J. Penman, *Biohistory* (Newcastle: Cambridge Scholars Publishing, 2015) for a full breakdown of the lemming cycle.

[169] S. N. DeWitte & J. W. Wood, "Selectivity of Black Death Mortality with Respect to Pre-Existing Health," *Proceedings of the National Academy of Sciences USA* 105 (5) (2008): 1436–41.

[170] K. B. McFarlane, *The Nobility of Later Medieval England* (Oxford: Clarendon Press, 1973), 150, 168–70

[171] G. Clark, *A Farewell to Alms: A Brief Economic History of the World* (Princeton, NJ: Princeton University Press, 2007), chapter three.

[172] P. Biller, "Birth Control in the West in the Thirteenth and Fourteenth Centuries" in Social History Society Newsletter (Spring 1980).

[173] E. A. Wrigley & R. S. Schofield, *The Population History of England 1541–1871* (London: Edward Arnold, 1981), 528–9.

[174] See Penman, *Biohistory* for further details.

[175] For fuller information on all cycles see ibid.

[176] For fuller information on all cycles see ibid.

[177] L. W. Pye, *Warlord Politics: Conflict and Coalition in the Modernization of Republican China* (New York: Praeger Publishers, 1971), 114–115.

[178] G. Reuter, "The German Religion of Duty," *The New York Times Current History* 1 (1914–5), 170–173.

[179] J. M. Tanner, 1962, http://www.breastcancerfund.org/assets/pdfs/publications/falling-age-of-puberty. pdf (accessed September 6, 2014); P. E. Brown, "The Age at Menarche," *British Journal of Preventive & Social Medicine* 20 (1) (1966): 9–14; M.A. Bellis, J. Downing, J.R. Ashton, "Adults at 12? Trends in puberty and their public health consequences", *Journal of Epidemiology and Community Health* 60 (2006): 910-11

(i) Figures for Germany have been taken from P.E. Brown due to superior detail compared to the *Journal of Epidemiology and Community Health* and Tanner, ibid.
(ii) Figures from Britain taken from P.E. Brown up to 1950, then filled in from Tanner in the later part of the twentieth century.
(iii) France was taken from P.E. Brown up until 1920 due to superior detail, then from the *Journal of Epidemiology and Community Health.*

[180] E. Jünger, *The Storm of Steel: From the Diary of a German Storm-troop Officer on the Western Front* (New York: Howard Fertig, Inc., 1996), 1.

[181] Ibid., 253–5.

[182] J. J. Spengler, *France Faces Depopulation* (New York: Greenwood Press, 1968), 53. B. R. Mitchell, *International Historical Statistics: Europe, 1750–2000* (Basingstoke, England: Palgrave Macmillan, 2003), 95.

[183] J.A. Mangan, "Duty unto Death: English Masculinity and Militarism in the Age of the New Imperialism," in *Tribal Identities: Nationalism, Europe, Sport*, edited by J. A. Mangan (Southgate, Frank Cass Publishers, 1996).

[184] Brian R. Mitchell, *International Historical Statistics: Africa, Asia & Oceania, 1750–2000* (London: Macmillan, 2003), 71–4; Statistical Handbook of Japan (English version). Edited by Statistical Research and Training Institute, published by Statistics Bureau of Japan, Ministry of Internal Affairs and Communications. Chapter Two: Population. http://www.stat.go.jp/english/data/handbook/c0117 .htm#c02 (accessed September 7, 2014).

[185] For full examples with statistics see J. Penman, *Biohistory* (Newcastle: Cambridge Scholars Publishing, 2015), chapter ten.

[186] R. Middleton & A. Lombard, *Colonial America, A History to 1763*, 4th Edition (United Kingdom: Wiley-Blackwell, 2011), 255.

[187] Ibid.

[188] M. R. Haines & R. H. Steckel, *A Population History of North America* (Cambridge: Cambridge University Press, 2000), 178.

[189] R. Raphael, *A People's History of the American Revolution: How Common People Shaped the Fight for Independence* (New York: New York Press, 2001), 47.

[190] Ibid.

[191] Ibid.

[192] A. Stephanson, *Manifest Destiny, American Expansion and the Empire of Right* (New York: Harper Collins, 2005), 42.

[193] Haines & Steckel, *A Population History of North America*, 315.

[194] Ibid., 50, 51.

[195] Ibid., 52

[196] Ibid.

[197] Yasukichi Yasuba, *Birth Rates of the White Population of the United States, 1800–1860: An Economic Study* (Baltimore: Johns Hopkins Press, 1962).

[198] Stephanson, *Manifest Destiny*, 64.

[199] C. Vann Woodward, "Introduction," in *Battle Cry of Freedom*, edited by James McPherson (Oxford: Oxford University Press, 2003), xix; Maris Vinovskis, *Toward a Social History of the American Civil War: Exploratory Essays* (Cambridge: Cambridge University Press, 1990), 7.

[200] George Edgar Turner, *Victory Rode the Rails: the Strategic Place of the Railroads in the Civil War* (1972); Steven G. Collins, "System in the South: John W. Mallet, Josiah Gorgas, and uniform production at the confederate ordnance department," *Technology and Culture* 40 (3) (1999): 517–544; Frank E. Vandiver, "Texas and the Confederate Army's Meat Problem," *Southwestern Historical Quarterly* 47 (3) (1944): 225–233; Larry J. Daniel, *Soldiering in the Army of Tennessee: A Portrait of Life in a Confederate Army* (Chapel Hill: University of North Carolina Press, 2003), chapter four.

[201] D. Kyvig, *Daily Life in the United States, 1920—1940. How Americans Lived through the Roaring Twenties and the Great Depression* (Chicago: Ivan R. Dee, 2002), 214–215.

[202] Jerome Blum, Rondo Cameron & Thomas G. Barnes, *The European World: a History* (1970), 885.

[203] Department of Health and Human Services, National Center for Health Statistics, www.dhhs.gov
National Vital Statistics Reports 61 (1)
"U.S. Census Bureau Announces 2010 Census Population Counts—Apportionment Counts Delivered to President" (Press release). *United States Census Bureau.* December 21, 2010

[204] T. H. Holmes & R. H. Rahe, "The Social Readjustment Rating Scale," *Journal of Psychosomatic Research* 11 (1967): 213–218.

[205] A. P. Blaszczynski & N. McConaghy, "Anxiety and/or Depression in the Pathogenesis of Addictive Gambling," *Substance Abuse* 24 (4) (1989): 337–350; G. J. Coman, G. D. Burrows & B. J. Evans, "Stress and Anxiety as Factors in the Onset of Problem Gambling: Implications for Treatment," *Stress Medicine* 13 (4) (1997): 235–44; A. C. Miu, R. M. Heilman & D. Houser, "Anxiety Impairs Decision-Making: Psychophysiological Evidence from an Iowa Gambling Task,"

Biological Psychology 77 (3) (2008): 353–8; A. P. Blaszczynski, A. C. Wilson & N. McConaghy, "Sensation Seeking and Pathological Gambling," *British Journal of Addiction* 81(1) (1986): 113–7.

[206] R. J. Evans, *The Third Reich in Power* (New York: The Penguin Press, 2005), 23.

[207] A. Hitler, *Mein Kampf*, http://www.greatwar.nl/books/meinkampf/meinkampf.pdf (accessed September 3, 2014).

[208] E. Lucie-Smith, *Art of the 1930s: the Age of Anxiety* (London: Weidenfeld & Nicolson, 1985).

[209] O. Semoyonova Tian-Shanskaia, *Village Life in Late Tsarist Russia* (Bloomington: Indiana University Press, 1993), 25, 27.

[210] Ibid., 6, 7, 30.

[211] Ibid., 50, 51.

[212] R. Bova, *Russia and Western Civilization* (New York: M.E. Sharpe, 2003), 271.

[213] Evans, *The Third Reich in Power*, 66.

[214] J. Coates, *The Hour Between Dog and Wolf: Risk-taking, Gut Feelings and the Biology of Boom and Bust* (London: Harper Collins, 2012).

[215] "Briefing: The Staid Young," *The Economist*, July 12, 2014.

[216] Jack Goldstone observes the link between population growth and revolution, though providing a different explanation; J. Goldstone, *Revolution and Rebellion in the Early Modern World* (Oakland: University of California Press, 1991).

[217] N. M. Chagnon, *The Yanomamo* (Fort Worth: Harcourt Brace College Publishers, 1997), 89.

[218] G. B. Grinnell, *The Cheyenne Indians: Their History and Lifeways* (Bloomington, Ind.: World Wisdom, 2008); G. E. Hyde, *Red Cloud's Folk: A History of the Oglala Sioux Indians* (Norman: University of Oklahoma Press, 1937); G. Gibbon, *The Sioux: The Dakota and Lakota Nations* (Oxford: Blackwell Publishers, 2003), 4.

[219] Gibbon, *The Sioux*, 74; Royal B. Hassrick, *The Sioux: Life and Customs of a Warrior Society* (Norman: University of Oklahoma Press, 2012), 8, 62–78.

[220] Ibid., 73–74.

[221] M. Mead, *Sex and Temperament in Three Primitive Societies* (New York: Harper Perennial, 2001), 232; D. E. Brown, *Human Universals* (New York: McGraw Hill, 1991); Education Portal, "Tchambuli Tribe: Culture, Gender Roles & Lesson" http://education-portal.com/academy/lesson/tchambuli-tribe-culture-gender-roles-lesson.html#lesson (accessed September 3, 2014).

[222] B. Sykes, *Saxons, Vikings, and Celts: The Genetic Roots of Britain and Ireland* (New York: WW Norton & Company, 2006).

[223] H. McKillop, *The Ancient Maya: New Perspectives* (New York: Norton, 2004).

[224] Other notable works include the Pillow Book of Sei Shonagon; M. Shikibu, *The Tale of Genji*, edited and translated by Royall Tyler (New York and London: Penguin Classic, 2002); W. Aston, *A History of Japanese Literature* (London: Heineman, 1907); D. Keene, *The Pleasures of Japanese Literature* (New York: Columbia UP, 1988).

[225] I. Morris, *The World of the Shining Prince* (Oxford: Oxford University Press, 1964), 26, 80, 122, 143–147, 170, 177–196, 211–212.

[226] H.-T. Fei, "Peasantry and Gentry: An Interpretation of Chinese Social Structure and its Changes," in *Social Structure and Personality*, edited by Y. A. Cohen, 24–35 (London: Holt, Rinehart and Winston, 1961); P.-t. Ho, *The Ladder of Success in Imperial China* (New York: Columbia University Press, 1962), 129–145, 157–160, 166; E. O. Reischauer & J. K. Fairbank, *East Asia: The Great Tradition* (Boston: Houghton Mifflin, 1960), 187, 223–228; F. Hsiao-Tung, *China's Gentry: Essays in Rural-Urban Relations*, edited by M. Park, , 205–206, 246–247, 270–272 (Chicago: University of Chicago Press, 1953).

[227] R. R. Grinker, "The Poor Rich: The Children of the Super-Rich," *American Journal of Psychiatry* 135 (1978): 913–916; R. R. Grinker, "The Poor Rich," *Psychology Today* (1977): 74–6, 81.

[228] Grinker, "The Poor Rich."

[229] Louis Henry, "The Population of France in the Eighteenth Century," in *Population in History*, edited by D. V. Glass & D. E. C. Eversley (London: Edward Arnold, 1965), 443–4; B. R. Mitchell, *International Historical Statistics: Europe, 1750–2000* (Basingstoke: Palgrave Macmillan, 2003), 95–6; Council of Europe 2002, "Recent Demographic Developments in Europe." Belgium: Council of Europe Publishing, December 2002, 34.

[230] Mitchell, *International Historical Statistics*, 100–116; Council of Europe, "Recent Demographic Developments in Europe," 34.

[231] H. L. Matthews, *Castro: A Political Biography* (London: Penguin, 1969), 67, 129; S. Diaz-Briquets & L. Perez, "Fertility Decline in Cuba: A Socioeconomic Interpretation," *Population and Development Review* 8 (3) (1982): 513–37; L. A. Perez, *Lords of the Mountain: Social Banditry and Peasant Protest in Cuba: 1878–1918* (Pittsburg: University of Pittsburg Press, 1989), 24–25.

[232] E. Gonzales, "After Fidel: Political Succession in Cuba," in *Cuban Communism*, edited by I. Horowitz, 499 (New Brunswick: Transaction, 1988).

[233] Y. Nakash, *The Shi'is of Iraq* (Princeton University Press, 2003), 136

[234] T. J. Cornell, *The Beginnings of Rome. Italy and Rome from the Bronze Age to the Punic Wars (c. 1000–264 BC)* (London: Routledge, 2004), 272–292.

[235] Plutarch, "Marcus Cato," in *Plutarch: Lives of the Noble Grecians and Romans*, edited by A. H. Clough, Project Gutenberg, (1996), http://www.gutenberg.org/cache/epub/674/pg674.html (accessed September 3, 2014).

[236] W.W. Fowler, *Social Life at Rome in the Age of Cicero* (London: MacMillan,1908), 204–546.

[237] Sallust, *The War with Cataline* http://penelope.uchicago.edu/Thayer/E/Roman/Texts/Sallust/Bellum_Catilinae*.html (accessed September 7, 2014).

[238] Fowler, Social Life at Rome, 101.

[239] A. H. McDonald, *Republican Rome* (London: Thames and Hudson, 1966), 76.

[240] J. Carcopino, *Daily Life in Ancient Rome: The People and the City at the Height of the Empire* (Harmondsworth: Penguin, 1956), 97.

[241] R. M. Geer & H. N. Couch, *Rome. Classical Civilization* (Englewood Cliffs, Prentice-Hall, 1950), 90

[242] Fowler, Social Life at Rome, 36

[243] J. Carcopino, *Daily Life in Ancient Rome: The People and the City at the Height of the Empire* (Harmondsworth: Penguin, 1956), 96.

[244] H. H. Scullard, A History of Rome from 133 B.C to 68 A.D (New York, Routledge, 1982), 33.

[245] Ibid., 111.

[246] Plutarch, "Life of Antony," in *Plutarch's Lives* vol. 2 (Digireads, 2009), http://books.google.com.au/books?id=H5lcpDJ_u44C&pg=PA355&lpg=PA355&dq=Antony.#v=onepage&q=Antony.&f=false (accessed September 3, 2014).

[247] C. Dio, *History*, translated by E. Carey (Cambridge: Harvard University Press, 1990), 4–5.

[248] Tacitus (c. 55–120), Annals, 14:15

[249] Geer & Couch, *Rome*, 373; P. Garnsey, "Child Rearing in Ancient Italy," in *The Family in Italy: From Antiquity to the Present*, edited by D. I. Kertzer & R. P. Saller, 60, 48–65 (New Haven and London: Yale University Press, 1991).

[250] Tacitus, Dialogus de Oratoribus 28, 29, quoted in A. R. Colón & P. A. Colón, *A History of Children: A Socio-Cultural Survey across Millennia* (Westport: Greenwood Press, 2001), 84–5.

[251] E. T. Salmon, A History of the Roman World from 30 B.C. to 138 A. D. (London: Methuen and Co., 1968), 253.

[252] E. Badian, *Foreign Clientelae, 264–70 BC* (Oxford: Clarendon Press, 1958), 197–211.

[253] W. R. Halliday, *Pagan Background of Early Christianity* (Liverpool: Liverpool University Press, 1925), 48.

[254] H. M. D. Parker and B. H. Warmington, *A History of the Roman World from A.D. 138–337* (London, Methuen, 1958), 69.

[255] Dio, *History*, 273.

[256] F. Lot, *The End of the Ancient World* (London: Routledge and Kegan Paul, 1953), 128–30.

[257] H. M. D. Parker and B. H. Warmington, *A History of the Roman World from A.D. 138–337* (London, Methuen, 1958), 240.

[258] R. Duncan-Jones, *The Economy of the Roman Empire* (Cambridge: Cambridge University Press, 1974), 318.

[259] Plini the Younger, *Complete Letters* (Oxford world's classics), translated by P. G. Walsh, (Oxford: Oxford University Press, 2006), 96.

[260] F. Lot, *The End of the Ancient World* (London, Routledge & Kegan Paul, 1953), 67.

[261] A. H. M. Jones, *The Late Roman Empire: 284–602* (Oxford: Basil Blackwell, 1973) Vol II, 796, 798, 803–8.

[262] S. Hong, J. Candelone, C. C. Patterson, C. F. Boutron, "History of Ancient Copper Smelting Pollution During Roman and Medieval Times Recorded in Greenland Ice," *Science* (272) (1996): 246–249; D. M. Settle, C. C. Patterson,

"Lead in Albacore: Guide to Lead Pollution in Americans," *Science* (207) (1980): 1167–1176.

[263] Parker & Warmington, *A History of the Roman World from A.D. 138–337* (London, Methuen, 1958) 39, 165.

[264] J. Lindsay, *Daily Life in Roman Egypt* (London, Frederick Müller, 1963), xix–xxi.

[265] Parker & Warmington, *A History of the Roman World from A.D. 138–337* (London, Methuen, 1958), 282–3.

[266] J. B. Bury, H. M. Gwatkin, J. P. Whitney, J. R. Tanner, C.W. Previte-Orton & Z. N. Brooke, *The Cambridge Medieval History. The Rise of Sacarens and the foundation of the Western Empire* Vol. 2 (New York: Macmillan, 1991) 40–41, 548–51.

[267] F. Lot, *The End of the Ancient World*, (London, Routledge & Kegan Paul, 1958), 151.

[268] C. Bailey, *Phases in the Religion of Ancient Rome* (Oxford: Oxford University Press, 1932), 246–8.

[269] D. C. A. Shotter, *Nero* (London, New York: Routledge, 2005), 23–26.

[270] M. Grant, *Nero, Emperor in Revolt* (New York: American Heritage Press, 1970), 151–2, 163.

[271] D.R. Dudley, *The World of Tacitus* (London: Secker and Warburg, 1968), 123; Grant, *Nero, Emperor in Revolt*, 212.

[272] Halliday, *Pagan Background of Early Christianity*, 127.

[273] S. Perowne, *Hadrian* (London: Hodder and Stoughton, 1960), 77–8.

[274] Perowne, *Hadrian*, 49–50.

[275] Geer & Couch, *Rome*, 372.

[276] Ibid., 367; Carcopino, *Daily Life in Ancient Rome*, 105.

[277] A.H.M. Jones, *The Late Roman Empire: 284–602* (London, Basil Blackwell, 1966), 978–9.

[278] R. MacMullen, *The Corruption and Decline of Rome* (New Haven: Yale University Press, 1988), 60–63.

[279] Parker & Warmington, *A History of the Roman World from A.D. 138–337* (London, Methuen, 1958), 29–35, 89–91.

[280] A.H.M. Jones, *The Late Roman Empire: 284–602*, (Oxford, Basil Blackwell, 1973) Vol II, 363.

[281] P. Charanis, "Observations on the Demography of the Byzantine Empire." Thirteenth International Congress of Byzantine Studies, Main Papers XIV, (Oxford, 1966), 11–12.

[282] A.H.M. Jones, *The Late Roman Empire: 284–602*, (Oxford, Basil Blackwell, 1973) Vol II, 1064.

[283] M.M. Sage, *Warfare in Ancient Greece: A Sourcebook* (London, Routledge, 1996): 281

[284] G. Cochran, J. Hardy & H. Harpending, "Natural History of Ashkenazi Intelligence," *Journal of Biosocial Science* 38 (5) (2006): 659–93.

[285] G. Cochran & H. Harpending, *The 10,000 Year Explosion: How Civilization Accelerated Human Evolution* (New York: Basic Books, 2009).

[286] Zoran Pavlovic, *Italy* (Philadelphia: Chelsea House, 2004), 49; "Birth and Fertility Rates among the Resident Population," Istituto Nazionale di Statistica (*Istat*), November 14, 2012. Statistics exclude non-Italian immigrants.

[287] Chinese historical information from: J. Keay, *China, a History* (London: Harper Collins, 2008), 64; E. O. Reischauer & J. K. Fairbank, *East Asia: The Great Tradition* (Boston, Houghton Mifflin, 1960).

[288] Keay, *China, a History*, 64.

[289] P. Olivelle, *Between the Empires: Society in India 300 BC to 400 AD* (Oxford: Oxford University Press, 2006), 36.

[290] B. B. Whiting, (ed.), *Six Cultures: Studies of Child Rearing* (New York and London: John Wiley and Sons, 1963): 513–518.

[291] S. C. Dube, *Indian Village* (London: Routledge & Kegan Paul, 1965).

[292] Plato, Laws, Book 7, http://hoodmuseum.dartmouth.edu/exhibitions/coa/ch_discipline.html (accessed September 3, 2014).

[293] Aristotle, *The Basic Works of Aristotle: Politics*, edited by R. McKeon (New York: The Modern Library, 1941), book 7, chapter 16.

[294] Ibid., book 7, chapter 17.

[295] Cyril Mango, *Byzantium: The Empire of New Rome* (New York: Scribner, 1980).

[296] 2010 ACS 1-year estimates. http://factfinder2.census.gov/faces/nav/jsf/pages/searchresults.xhtml?refresh=t (accessed August 14, 2013).

[297] J. G. Macqueen, *Babylon* (London: Robert Hale, 1964), 73–5.

[298] S. W. Bauer, *The History of the Ancient World: From the Earliest Accounts to the Fall of Rome* (New York & London: WW Norton, 2007), 133–35.

[299] Ibid., 128.

[300] L. Woolley, *The Sumerians* (Oxford: The Clarendon Press, 1928). xii, 75; M. Brosius, *The Persians: An Introduction* (Abington: Routledge, 2006), 88, 93, 143, 153–157; J. Haywood, *The Penguin Historical Atlas of Ancient Civilizations* (London: Penguin Books Ltd., 2005), 22–53; P. K. O'Brien, *Atlas of World History* (London: Philip's, 2007), 28, 39, 41–43, 53; H. Saggs, *Civilization before Greece and Rome* (New Haven: Yale University Press, 1989), 17–18.

[301] E. Lipinski, *The Aramaeans: their Ancient History, Culture, Religion* (Leuven: Peeters, 2000).

[302] The Mongols also adopted new military and political structures in the period leading up to their great expansion, presumably also the result of a G period. For further information see J. Penman, *Biohistory* (Newcastle: Cambridge Scholars Publishing, 2015), Chapter 15.

[303] A. Barlas, *Believing women in Islam* (Austin, University of Texas Press, 2002) K. Armstrong, *Muhammad: A Biography of the Prophet,* (New York, Harper Collins, 1992)

[304] US Department of State, "Lebanon," International Religious Freedom Report 2010, http://www.state.gov/j/drl/rls/irf/2010/148830.htm (accessed September 3, 2014).

[305] S. P. Ramet, *Nationalism and Federalism in Yugoslavia, 1962–1991* (Bloomington, IN: Indiana University Press, 1992).

[306] F. Fukuyama, *The End of History and the Last Man* (New York: Free Press, 1992).

[307] R. Hood & C. Hoyle, *The Death Penalty: a Worldwide Perspective*, 4th ed. (Oxford: Oxford University Press, 2008), 11–15, 49–50, 354.

[308] Nederlands Juristenblad 496, March 20, 1920; Council of Europe. *Eliminating Corporal Punishment: a Human Rights Imperative for Europe's children* (Strasbourg, Council of Europe Publishing 2005); Article 47(3) of the School Education Act; D. Schumann, "Legislation and Liberalization: The Debate About Corporal Punishment in Schools in Post-war West Germany, 1945—1975," *German History* 25 (2) (2007): 192–218.

[309] M. Pate & L. A. Gould, *Corporal Punishment around the World* (Santa Barbara: Praeger, 2012), 81–2.

[310] Ibid., 83.

[311] Teachers, "Women's Suffrage," http://www.scholastic.com/teachers/article/womenx2019s-suffrage (accessed September 3, 2014).

[312] Ibid.

[313] Council of Europe, Recent Demographic Developments in Europe, Demographic Yearbook, (2003), http://www.rand.org/pubs/research_briefs/RB9126/index1.html (accessed September 3, 2014).

[314] CIA World Factbook.

[315] Statistics Bureau of Japan Population by sex, population increase and population density. http://www.stat.go.jp/english/data/chouki/02.htm (accessed September 3, 2014); National Institute of Population and Social Security Research (December 2006). Population projections for Japan: 2006–2055, http://www.ipss.go.jp/index-e.asp (accessed September 3, 2014); *The Japanese Journal of Population* 6 (1) (2008): 76–114.

[316] Mark Adomanis, "'Dying' Russia's birth-Rate is Now Higher than America's," *Forbes* 4, November 2014.

[317] C. Kingsley, *Miscellanies*, Vol. II, (1863), 364.

[318] O. Bennett, *Narratives of Decline in the Postmodern World* (Edinburgh: Edinburgh University Press, 2001).

[319] Glenn Ricketts, Peter W. Wood, Stephen H. Balch & Ashley Thorne, "The Vanishing West: 1964–2010. The Disappearance of Western Civilization from the American Undergraduate Curriculum," *National Association of Scholars*, May 2011.

[320] GovToday, "New Report Reveals Sexual Behaviour across the Different Age Groups," (2011), http://www.govtoday.co.uk/health/44-public-health/9286-new-report-reveals-sexual-behaviour-across-the-different-age-groups (accessed September 3, 2014).

[321] Figures cover the entire US population, including minority groups: Ranking America, "The U.S. Ranks 13th in Age of First Sex," 2009, http://rankingamerica.wordpress.com/2009/01/28/the-us-ranks-13th-in-age-of-first-sex/ (accessed September 3, 2014); Real Clear Politics, "Virginity Rising, (2011), http://www.realclearpolitics.com/articles/2011/03/10/virginity_rising_109173.html (accessed September 3, 2014).

[322] Tanner (1962)
http://www.breastcancerfund.org/assets/pdfs/publications/falling-age-of-puberty
.pdf (accessed September 3, 2014); *Journal of Epidemiol Community Health* 60
(11) (2006): 910–911; P. E. Brown, "The Age at Menarche," *British Journal of
Preventive & Social Medicine* 20 (1) (1966): 9–14. Notes: (i) figures for Germany
and have been taken from this source due to superior detail compared to the
Journal of Epidemiol Community Health and Tanner; (ii) figures from Britain
taken from this source up to 1950, then filled in from Tanner in the later part of the
twentieth century.
[323] C. G. Brown, *Religion and Society in 20th-century Britain* (London: Longman,
2006), 32.
[324] B. Stevenson & J. Wolfers, "Marriage and Divorce: Changes and their Driving
Forces," *Journal of Economic Perspectives* 21 (2) (2007): 27–52.
[325] Selwyn Duke, "Vladimir Putin Caesar and Our Great Geo-political Turning
Point," *American Thinker*, April 25, 2014.
[326] Internet FAQ Archive, "Parenting," (2008), http://www.faqs.org/childhood/Me-
Pa/Parenting.html (accessed September 3, 2014).
[327] James E. Block, *The Crucible of Consent: American Child Rearing and the
Forging of Liberal Society* (Cambridge, MA: Harvard University Press, 2012).
[328] C. Smith, H. D. Christoffersen & P. S. Herzog, *Lost in Translation: The Dark
Side of Emerging Adulthood* (New York: Oxford University Press USA, 2011);
See also Theodore Dalrymple, *Life at the Bottom: The Worldview That Makes the
Underclass* (Rugby: Monday Books, 2010); Theodore Dalrymple, *Our Culture,
What's Left of It* (Rugby: Monday Books, 2010).
[329] US Government Spending, usgovernmentspending.com (accessed September 7,
2014).
[330] T. P. Jeffrey, "Census: 49% of Americans Get Gov't Benefits; 82M in Households
on Medicaid" *cnsnews.com* October 23, 2013 (accessed September 9, 2014).
[331] US Government Spending, http://www.usgovernmentspending.com/us_20th
_century_chart.html (accessed September 7, 2014).
[332] US Government Spending, http://www.usgovernmentspending.com/us_20th
_century_chart.html (accessed September 3, 2014).
[333] William A. Sundstrom, tables Ba4568 and Ba4589; Kendrick 1961, table D-10,
in "A Century of Work and Leisure," V. A. Ramey & N. Francis (2006),
http://weber.ucsd.edu/~vramey/research/Century_Published.pdf (accessed September
3, 2014).
[334] "Mind the gap", *The Economist*, April 11-17, 2015
[335] C. Murray, *Coming Apart: The State of White America, 1960–2010* (New York:
Crown Forum, 2012), 169.
[336] AMECO, US Fed Budget; IMF.
[337] B. H. Bunch & A. Hellemans, *The History of Science and Technology: A
Browser's Guide to the Great Discoveries, Inventions, and the People who made
them from the Dawn of Time to Today* (New York: Houghton Mifflin Harcourt,
2004); J. Huebner, "A Possible Declining Trend for Worldwide Innovation,"
Technological Forecasting and Social Change 72 (8) (2005): 980–6.

[338] L. Ward, "Physics Degree Courses Axed as Demand Slumps," *The Independent*, (January 23, 1997), http://www.independent.co.uk/news/uk/politics/physics-degree-courses-axed-as-demand-slumps-1284606.html (accessed September 3, 2007).
[339] "A is for algorithm," *The Economist*, (April 26, 2014).
[340] National Science Foundation/Division of Science Resources Statistics, Science and Engineering Doctorate Awards: 2000, Detailed Statistical Tables, NSF 02-305 (Arlington, VA, 2001).
[341] D. Mattei, "The Decline of Traditional Values in Western Europe," *International Journal of Comparative Sociology* 39 (1) (1998): 77–90.
[342] G. Koritzky, E. Yechim, I. Bukay & U. Milman, "Obesity and Risk Taking. A Male Phenomenon," *Appetite* 59 (2) (2012): 289–97; C. Nederkoorn, F. T. Y. Smulders, R. C. Havermans, A. Roefs & A. Jansen *Appetite* 47 (2) (2006): 253–256; A. Drewnowski & S. E. Specter "Poverty and Obesity: the Role of Energy Density and Energy Cost," *The American Journal of Clinical Nutrition* 79 (1) (2004): 6–16.
[343] E. Carlsen, A. Giwercman, N. Keiding & N. E. Skakkebæk, "Evidence for Decreasing Quality of Semen During past 50 Years," *British Medical Journal* 305 (6854) (1992): 609; J. Raloff, "That Feminine Touch: Are Men Suffering from Prenatal or Childhood Exposure to 'Hormonal' Toxicants?" *Science News* 145 (1994): 56–8; G. R. Bentley, "Environmental Pollutants and Fertility," in *Infertility in the Modern World: Present and Future Prospects*, edited by G. R. Bentley & C. G. N. Mascle-Taylor (New York: Cambridge University Press, 2000); J. A. Saidi, D. T. Chang, E. Goluboff, T., E. Bagiella, G. Olsen & H. Fisch, "Declining Sperm Counts in the United States? A Critical Review," *The Journal of Urology* 161 (2) (1999): 460–2; J. Auger, J. M. Kunstmann, F. Czyglik & P. Jouannet, "Decline in Semen Quality among Fertile Men in Paris during the Past 20 Years," *New England Journal of Medicine* 332 (5) (1995): 281–5; M. Rolland, J. Le Moal, V. Wagner, D. Royère & J. De Mouzon, "Decline in Semen Concentration and Morphology in a Sample of 26,609 Men Close to General Population between 1989 and 2005 in France," *Human Reproduction* 28 (2) (2013): 462–70.
[344] T. G. Travison, A. B. Araujo, A. B. O'Donnell, V. Kupelian & J. B. McKinlay, "A Population-Level Decline in Serum Testosterone Levels in American Men," *The Journal of Clinical Endocrinology & Metabolism* 92 (1) (2007): 196–202.
[345] Barry R. Bloom, *Tuberculosis: Pathogenesis, Protection, and Control* (Washington, D.C., ASM Press, 1994).
[346] L. G. Wilson, "The Historical Decline of Tuberculosis in Europe and America: its Causes and Significance," *Journal of the History of Medical and Allied Science* 45 (3) (1990): 366–396.
[347] World Health Organization (2011). "The Sixteenth Global Report on Tuberculosis."
[348] N. Bhatti, M. R. Law, J. K. Morris, R. Halliday & J. Moore-Gillon, "Increasing Incidence of Tuberculosis in England and Wales: a Study of the Likely Causes," *British Medical Journal* 310 (6985) (1995): 967–989; P. Davies, "The Worldwide Increase in Tuberculosis: how Demographic Changes, HIV Infection and Increasing

Numbers in Poverty are Increasing Tuberculosis," *Annals of Medicine* 35 (4) (2003): 235–243; R.A. Weiss & A. J. McMichael, "Social and Environmental Risk Factors in the Emergence of Infectious Disease," *Nature Medicine* 10 (2004): 570–576.

[349] Samuel H. Williamson, "What Was the U.S. GDP Then?" *Measuring Worth*, 2014.

[350] P. B. Kaplowitz & S. E. Oberfield, "Reexamination of the Age Limit for Defining when Puberty is Precocious in Girls in the United States: Implications for Evaluation and Treatment," *Pediatrics* 104 (4): (1999): 936–41; P. B. Kaplowitz, E. J. Slora, R. C. Wasserman, S. E. Pedlow & M. E. Herman-Giddens, "Earlier Onset of Puberty in Girls: Relation to Increased Body Mass Index and Race," *Pediatrics* 108 (2) (2001): 347–53; E. J. Susman, L. D. Dorn & V. L. Schiefelbein, "Puberty, Sexuality, and Health," *Comprehensive Handbook of Psychology*, edited by M. A. Lerner, M. A. Easterbrooks & J. Mistry (New York: Wiley, 2003).

[351] Murray, *Coming Apart*, 220.

[352] World Top Incomes Database, 2012. Missing values interpolated using 5% and top 1% series.

[353] D. Flavelle, "Why the Gap between Rich and Poor in Canada keeps Growing," The Star, http://www.thestar.com/business/article/1097055--why-the-gap-between-rich-and-poor-in-canada-keeps-growing (accessed September 3, 2014)."What do you do when you reach the top?," The Economist, (November 9, 2011), http://www.economist.com/node/21538104 (accessed September 7, 2014); Central Intelligence Agency, "The World Factbook"; Mason, "Britain: Income Inequality at Record High."

[354] Still unpublished studies suggest that early CR increases sperm count in rats.

INDEX